Practical
Math

GRADE 6
Advanced Version

Kwang S. Ko, Ph.D.

1 *A SELF-STUDY GUIDE*
2 *EXERCISES*
3 *SELF-TESTS*
4 *SOLVING PROBLEMS*
5 *FULL ANSWER KEY*

CONTENTS

CHAPTER 1
Number Sense

In this chapter, you will learn number sense in order to compute addition, subtraction, multiplication, and division expressed with two negative numbers and mixed negative numbers. Absolute value, properties of equalities, and properties of real numbers will be taught as well.

1. Adding, Subtracting, Multiplying, and Dividing Rational Numbers

1–1. Table 1. Operations can affect your answer.

In addition:	
$(-) + (-) = -$	$(-) + (+) = +$ or $-$
$(+) + (+) = +$	$(+) + (-) = +$ or $-$
In subtraction:	
$(-) - (+) = -$	$(-) - (-) = +$ or $-$
$(+) - (-) = +$	$(+) - (+) = +$ or $-$
In multiplication:	
$(-) \times (-) = +$	$(-) \times (+) = -$
$(+) \times (+) = +$	$(+) \times (-) = -$
In division:	
$(-) \div (-) = +$	$(-) \div (+) = -$
$(+) \div (+) = +$	$(+) \div (-) = -$

1–2. Find the value of $(-2) + (-1)$.

SOLUTION

In the expression, both numbers are negative. Using a number line, start at "–2", and then move one place to the left.

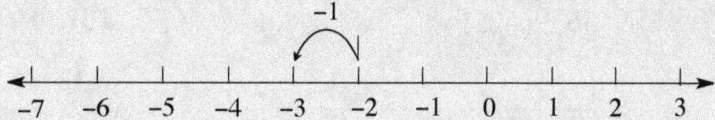

So the value in the expression is –3. Table 1 in the addition shown that is $(-) + (-) = -$.

Quick Exercises 1 Solve each expression.

1) $9 + (-3)$

2) $-3 + 5$

1-3. Find the value of $3 - (-5)$.

> **SOLUTION**
>
> In the expression, one number is positive and the other is negative. Start at 3 on the number line and as $-(-5) = +5$, moves 5 places to the right.
>
>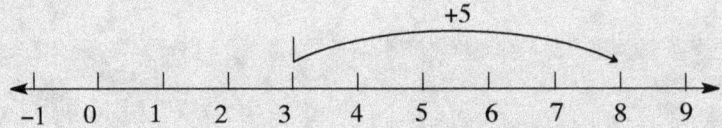
>
> So the value in the expression is 8. Table 1 shows that is $(+) - (-) = +$.

Quick Exercises 2   Solve each expression.

1) $(4) - (-3)$

2) $(-3) - (7)$

Exercises 1   Find the sum of the expressions.

1) $2 + (-2)$

2) $5 + (-7)$

3) $6 + (-2)$

4) $(-6) + 2$

5) $(-\$18) + \$36$

6) $\$12.25 + (-\$35.75)$

7) $(-25) + (-17)$

8) $(-\$22) + (-\$41)$

9) $36 + (-75)$

10) $19 + (-5)$

11) $9 + (-24)$

12) $4 + (-18)$

13) $7 + (-2)$

14) $(-32) + 17$

Exercises 2   Find the difference of the expressions.

1) $3 - (-14)$

2) $\$23 - (-\$15)$

3) $19 - (-21)$

4) $(-7) - 9$

5) $(-12) - 6$

6) $(-42) - 29$

7) $(-\$8.00) - (-\$15.00)$

8) $(-\$25) - (-\$18)$

9) $(-12) - (-21)$

10) $(-5) - (-8)$

11) $(-14) - 9$

12) $(-27) - (-15)$

Exercises 3   Find the value of Δ.

1) $(-\$60.00) + \Delta = \$7.00$

2) $\Delta + (-0.64) = -1.07$

3) $(-3.2) + \Delta = 5.1$

4) $-6 + \Delta = -17$

5) $\Delta - (-\$4.00) = \$2.00$

6) $(-6) - \Delta = (-3)$

7) $(-13) + \Delta = 18$

8) $\Delta - (-9) = (-14)$

9) $(-4) - \Delta = (-6)$

10) $\Delta + (-5) = (-5)$

11) $(-19) + \Delta = (-28)$

12) $\Delta - (-27) = 12$

1–4. Find the value of $(-3) \times (-5)$.

> **SOLUTION**
>
> In the expression, both numbers are negative. Table 1 shows that is $(-) \times (-) = +$. So the value of the expression is 15. In an expression, if both numbers are negative, then the answer is always positive. If both numbers are positive, then the answer is always positive. And if one of the numbers is negative, then the answer will be always negative.

Quick Exercises 3   Find the value of each expression.

1) $2 \times (-4)$ **2)** $(-3) \times 5$

1–5. Find the value of $(-16) \div (8)$.

> **SOLUTION**
>
> In the expression, one number is negative and the other number is positive. Table 1 shows that is $(-) \div (+) = -$. So, the value of the expression is -2.

Quick Exercises 4   Solve each expression.

1) $(-6) \div 2$ **2)** $(-25) \div (-5)$

Exercises 4   Find the product of the expressions.

1) $(-12) \times (-8)$ **2)** $0.4 \times (-0.2)$

3) $26 \times (-4)$ **4)** $(-5) \times (-14)$

5) $(-2.5) \times 0.2$ **6)** $(-2) \times \$18$

7) $1\frac{3}{5} \times (-1)$ **8)** $4\frac{2}{3} \times (\$12)$

Exercises 5   Find the quotient of the expressions.

1) $36 \div (-9)$

2) $72 \div (-4)$

3) $85 \div (-4)$

4) $(-225) \div 6$

5) $\$480 \div (-15)$

6) $\$462 \div (-8)$

7) $(-225) \div (-25)$

8) $(-48) \div (-2.4)$

9) $\$720 \div (-12)$

10) $144 \div (-16)$

11) $\Delta \div 3 = (-56)$

12) $(-28) \div (-14)$

13) $(-24) \div (-1)$

14) $(-6) \div 12$

Exercises 6   Find the value of Δ.

1) $2\frac{2}{5} \times \Delta = -36$

2) $\Delta \times (-16) = -4$

3) $\Delta \div 8 = (-12)$

4) $(-52) \div \Delta = (-4)$

5) $-1.36 \div \Delta = 0.5$

6) $\Delta \div (-6.8) = (-2)$

7) $1\frac{1}{6} \times \Delta = (-28)$

8) $2\frac{2}{3} \div \Delta = 8$

9) $\Delta \div 2\frac{1}{2} = (-8)$

10) $\Delta \times (-1.5) = (-6)$

2. Ratios

1–6. Express as a ratio.

                a. 1 to 5   b. 4 feet to 7 feet

SOLUTION

A ratio is a comparison between two numbers. It can be written out as shown by the following: 1) using the phrase " a to b", 2) using a colon "a : b", 3) writing it out as a fraction " $\frac{a}{b}$ ".

a. "1 to 5" means $1 : 5 = \frac{1}{5}$.

b. "4 feet to 7 feet" means $4 : 7 = \frac{4}{7}$.

       * Ratio: a ratio is a comparison of two numbers.
       * Equivalent ratios: two ratios that are equal to each other.
       * Proportion: a proportion is an equation that states two ratios are the same.

Quick Exercises 5   Express as a ratio.

    1) 2 to 5   **2)** 12 to 36

1–7. Write the ratios in the statement as an equation and determine whether or not if they are a proportion. Explain.

                3 to 8 as 9 to 24

SOLUTION

A proportion is a statement that compares two ratios.

a. "3 to 8" means $3:8 = \frac{3}{8}$ and "9 to 24" means $9:24 = \frac{9}{24}$

So you can determine whether the statement is a proportion if the cross products of the ratios are equal to each other.

$$\frac{3}{8} \bowtie \frac{9}{24}$$

$(3)(24) = (9)(8)$ Cross Products Property.
$72 = 72$ Simplify.

The cross products are equal to each other. Therefore the statement is a proportion and the equation is $\frac{3}{8} = \frac{9}{24}$.

1–8. Write the ratios in the statement as an equation and determine whether or not if they are a proportion. Explain.

5 to 7 as 10 to 11

SOLUTION

"5 to 7" means 5 : 7 or $\dfrac{5}{7}$ and "10 to 11" means 10 : 11 or $\dfrac{10}{11}$.

So you can determine whether the statement is a proportion if the cross products are equal to each other.

$$\dfrac{5}{7} \bowtie \dfrac{10}{11}$$

$(5)(11) = (10)(7)$	Cross Products Property
$55 \neq 70$	Simplify.

The cross products are NOT equal to each other. So, the statement is NOT a proportion and the equation is $\dfrac{5}{7} \neq \dfrac{10}{11}$.

Quick Exercises 6   Write the ratios in the statement as an equation and determine whether or not if they are a proportion. Explain.

1) 2 to 5 as 6 to 15   **2)** 7 to 3 as 9 to 5

1–9. Express as a ratio.

4 g to 5 kg

SOLUTION

When comparing two measurements in a ratio, they should both have the same unit. Otherwise, if they don't match each other, they will have to be converted to the same unit.

"4 g to 5 kg" = 4 g to 5 kg × (1000g/1 kg)	Convert to like units.
= 4 g to 5000 g	Simplify. $\left(5\text{kg} \times \dfrac{1000\text{g}}{1\text{ kg}} = 5000\text{g}\right)$
= 4 : 5000	Write ratio.
$= \dfrac{4}{4} : \dfrac{5000}{4}$	Simplify.
= 1 : 1250	

Quick Exercises 7   Express as a ratio.

1) 140 g to 0.35 kg   **2)** 18 cm to 0.63 m

Exercises 7   Write the ratios in the statement as a proportion and determine if it is true. Explain.

1) 7 is to 8 as 28 is to 33.

2) 6 is to 10 as 4.2 is to 7.0.

Exercises 8   Write the ratios in the statement as a proportion and solve for x.

1) 11 to 48 as x to 22.

2) $x + 3$ to 12 as $2x$ to 6.

3) $(3x - 2)$ to 6 as 2 to 10.

4) $(2x + 5)$ to 5 as 15 to 10

Exercises 9   Simplify the ratios.
        \<See list of conversions on page 45>

1) $\dfrac{720 \text{ g}}{0.32 \text{ kg}}$

    hint: 1 kg = 1000 g

2) $\dfrac{3.5 \text{ m}}{600 \text{ cm}}$

    hint: 1 m = 100 cm

3) $\dfrac{6 \text{ lb}}{80 \text{ oz}}$

    hint: 1 pound (lb) = 16 ounces (oz)

4) $\dfrac{32 \text{ in}}{4 \text{ ft}}$

    hint: 1 foot = 12 inches (in)

5) $\dfrac{4 \text{ kg}}{2500 \text{ g}}$

6) $\dfrac{58 \text{ cm}}{0.6 \text{ m}}$

7) $\dfrac{10 \text{ yd}}{18 \text{ ft}}$

    hint: 1 yard = 3 feet

8) $\dfrac{820 \text{ m}}{1.2 \text{ km}}$

    hint: 1 km = 1000 m

9) $\dfrac{12 \text{ ft}}{64 \text{ in}}$

10) $\dfrac{8 \text{ ft}}{72 \text{ in}}$

11) $\dfrac{48 \text{ oz}}{3 \text{ lb}}$

12) $\dfrac{20 \text{ cm}}{1.5 \text{ m}}$

SELF-TEST

1. Which of the following can be written as the ratio of *a* and *b*?

    A. $\frac{a}{b}$, where b ≠ 0.

    B. $a : b$

    C. a to b

    D. All of the above.

2. Which of the following ratios is equivalent to 15 : 21?

    A. 7 to 5

    B. 12 to 18

    C. 5 to 7

    D. 40 to 50

3. Which of the following is a proportion?

    A. $\frac{1}{5} ? \frac{3}{15}$

    B. $\frac{2}{3} ? \frac{14}{21}$

    C. $\frac{1}{2} ? \frac{3}{6}$

    D. All of the above.

4. At the aquarium, the orcas are fed 20.5 kilograms of fish. The dolphins are fed 12250 grams as many than the orcas. What is the ratio of the amount of fish fed to the orcas and dolphins in kilograms? (1000 grams in 1 kilogram)

    A. 16.7 to 1

    B. 20.5 to 12250

    C. 1 to 16.7

    D. 12250 to 20.5

5. Lisa works at an ice cream store. She sold 54 ice cream cones. The next day, she sold 175 more ice cream cones than what she sold the previous day. What is the ratio of the number of ice cream cones sold on both days?

    A. 4.24 to 1

    B. 1 to 3.24

    C. 1 to 4.24

    D. 3.24 to 1

6. An ice cream store sold 64 vanilla cones and 47 strawberry cones. What is the ratio of the number of vanilla to strawberry ice cream cones?

    A. 1.36 to 1

    B. 1 to 1.36

    C. 64 to 111

    D. 47 to 111

7. A bakery store used 878 eggs in the morning to bake their products. The bakers used 42 eggs to make chocolate chip cookies and used 305 eggs to make cakes. What is the ratio of the number of eggs used to bake chocolate chip cookies and cakes?

    A. 20.9 to 1   **B.** 1 to 2.88

    C. 1 to 7.26   **D.** 7.26 to 1

8. Jane is sawing wood to make a shelf. She originally had a 14 ft board. After sawing a piece off, she now has a $9\frac{1}{6}$-inch board. What is the ratio of the original length of the board to the new length?

    A. 18.34 to 1   **B.** 14 to $9\frac{1}{6}$

    C. 1 to 18.34   **D.** $9\frac{1}{6}$ to 14

  * Use the information for Exercises **9-11**. Poppy is making a smoothie. She plans on putting 1/4 pounds of strawberries, 1/6 pounds of bananas, 1/3 pounds of raspberries, and 1/3 ounce of sugars in her smoothie.

9. What is the ratio of the number of strawberries to bananas in the smoothie?

    A. 3 to 4   **B.** 2 to 3

    C. 4 to 3   **D.** 3 to 2

10. What is the ratio of the number of strawberries to sugar in the smoothie?

    A. 3 to 4   **B.** 3 to 64

    C. 4 to 3   **D.** 64 to 3

11. Which of the following ratios is equivalent to 6 : 72?

    A. 12 to 1   **B.** 3 to 22

    C. 1 to 12   **D.** 2 to 27

3. Algebra: Ratios

1–10. Simplify the ratio.

$$\frac{12}{x} = \frac{2}{5}$$

SOLUTION

You can solve this problem in two ways. You can use the Cross Product Property to solve the equation.

$\dfrac{12}{x} = \dfrac{2}{5}$	Original proportion
$(12)(5) = 2x$	Cross Product Property
$60 = 2x$	Simplify.
$x = 30$	Divide each side by 2.

You can also use the Reciprocal Property as an alternative to solve the equation.

$\dfrac{12}{x} = \dfrac{2}{5}$	Original proportion
$\dfrac{x}{12} = \dfrac{5}{2}$	Reciprocal property
$x = (12)\dfrac{5}{2}$	Multiply both sides by 12.
$x = 30$	Simplify.

So the value of x is 30.

Quick Exercises 8   Solve each equation.

1) $\dfrac{x}{2} = \dfrac{3}{4}$

2) $\dfrac{x-2}{3} = \dfrac{3}{9}$

3) $\dfrac{6}{x} = \dfrac{2}{3}$

4) $\dfrac{1}{2x+1} = \dfrac{1}{3}$

Exercises 10 Solve each equation.

1) $\dfrac{a}{12} = \dfrac{22}{28}$

2) $\dfrac{m}{9} = \dfrac{7}{42}$

3) $\dfrac{a}{20} = \dfrac{8}{10}$

4) $\dfrac{2}{5} = \dfrac{8}{y}$

5) $\dfrac{5}{x} = \dfrac{1}{8}$

6) $\dfrac{9}{x} = \dfrac{1}{3}$

7) $\dfrac{2}{3x+4} = \dfrac{2}{7}$

8) $\dfrac{y+1}{3} = \dfrac{1}{9}$

9) $\dfrac{4}{3x-2} = \dfrac{1}{6}$

10) $\dfrac{2x-3}{9} = \dfrac{4}{12}$

11) $\dfrac{x-1}{3} = \dfrac{2}{3}$

12) $\dfrac{3}{x+2} = \dfrac{1}{4}$

Exercises 11 Write a proportion in order to find the value of x.

1) 20 miles traveled in 20 minutes
85 miles traveled in x minutes

2) 10 movie tickets for $130.00
4 movie ticket for $$x$

3) $27.60 for 6 pounds of T-bone steak
$$x$ for 2 pounds of T-bone steak

4) 12 water bottles for $18.00
4 water bottles for $$x$

5) 5 notebooks for $3.00
16 notebooks for $$x$

6) $6.75 for 8-pack AA battery
$$x$ for 24-pack AA battery

Exercises 12   Solve each problem using the given information.

1) What is the value of x if the ratios of each corresponding side length in the shape are in the extended ratio of 1: 8?

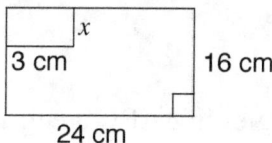

2) What is the value of x if the ratios of each side lengths in the shape below are in the extended ratio of 1: 12?

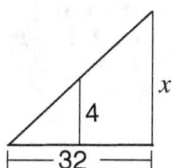

3) A table is set out for a buffet. On the table are plates of shrimp. Each plate has 152 pieces of shrimp. Which proportion could be set out to find A, the number of plates on the table, if there are 456 pieces of shrimp in total?

1. Which of the following is a true proportion?

   A. $\dfrac{16}{7} = \dfrac{7}{16}$ **B.** $\dfrac{13}{26} = \dfrac{39}{78}$

   C. $\dfrac{2}{5} = \dfrac{8}{21}$ **D.** $\dfrac{3}{5} = \dfrac{6}{5}$

2. The ratio of the lengths of $\triangle ABC$ are 4 : 7 : 5 and the shortest length AC is 19 in. Find the length of BC given that the perimeter of $\triangle ABC$ is 76 in^2. Round your answer to the nearest tenth.

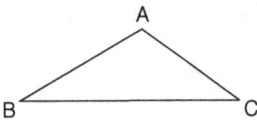

    A. 33.3 in.   **B.** 23.8 in.
    C. 42.8 in.   **D.** 38.0 in.

3. The ratios of the angles in ΔABC are 9: 23: 12. Find the measure of the biggest angle of ΔABC. Round your answer to the nearest tenth.

    A. 102.2°   **B.** 94.1°
    C. 49.0°   **D.** 36.8°

4. If the side lengths of ΔABC are 4 cm, 6 cm, and 8 cm and the ratio of the side lengths of ΔABC to ΔDEF is 2 : 7, which of the following is the perimeter of ΔDEF?

    A. 36 cm   **B.** 18 cm
    C. 126 cm   **D.** 63 cm

5. ΔABC is similar to ΔDEF and the ratio of the perimeter of ΔABC to ΔDEF is 7 : 8. Find the perimeter of ΔDEF if the perimeter of ΔABC is 59.8 in. Round your answer to the nearest tenth.

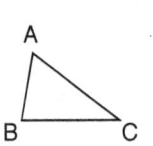

    A. 418.6 in.   **B.** 68.3 in.
    C. 52.3 in.   **D.** 478.4 in.

6. ΔABC is similar to ΔDEF. The ratio of the perimeters of ΔABC to ΔDEF is 13:7. Find the perimeter of ΔDEF if the perimeter of ΔABC is 140 in. Round your answer to the nearest tenth.

    A. 260.0 in.   **B.** 980 in.
    C. 75.4 in.   **D.** 1820 in.

7. ABCD is similar to EFGH. Find the length of CD.

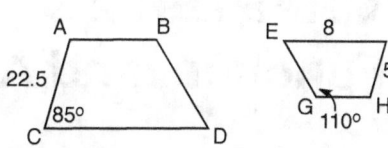

    A. 24   **B.** 28

    C. 32   **D.** 36

8. $\triangle$ABC is similar to $\triangle$DEF. Find the length of DE.

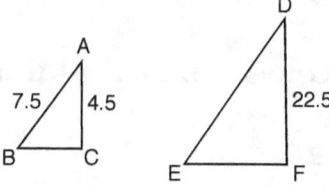

    A. 28.5   **B.** 32.5

    C. 37.5   **D.** 39.5

9. A miser is organizing his coins into several bags. If he has 6 bags filled with 228 coins, how many bags would he fill if he had 418 coins, given that each bag has an equal number of coins?

    A. 9   **B.** 10

    C. 12   **D.** 11

10. Given that 11 people bought movie tickets that cost $187 altogether, how many tickets were bought for $594?

    A. 28   **B.** 3

    C. 35   **D.** 4

11. If a car travels 48 miles in 40 minutes, how long will it take to travel 250 miles?

    A. 3h 47m   **B.** 3h 10m

    C. 5h   **D.** 5h 20

CHAPTER 2
Patterns, Function, and Algebra

In this chapter, you will identify number sense based on the number system, algebraic expression, and order of operations. Also you will learn about graphing functions, plotting the coordinate, relating function tables, and finding the equations of function tables.

1. Solving Equation with One Variable: Addition and Subtraction

2-1. Solve the equation.

$$-4 + \frac{x}{3} = 5$$

SOLUTION

a) First, look at the equation. In the equation, x represents the unknown value and is called the variable.

$-4 + \dfrac{x}{3} = 5$	Original equation
$(4) - 4 + \dfrac{x}{3} = 5 + (4)$	Add (4) to both sides.
$\dfrac{x}{3} = 9$	Simplify. $4 - 4 = 0$, $5 + 4 = 9$
$3 \times \dfrac{x}{3} = 9 \times 3$	Multiply each side by 3.
$x = 27$	Simplify. $3 \times \dfrac{x}{3} = x$. $9 \times 3 = 27$

So, x represents 27.

Check the solution.

$-4 + \dfrac{x}{3} = 5$	Original equation
$-4 + \dfrac{27}{3} = 5$	Substitute x with 27.
$-4 + 9 = 5$	Simplify the fraction. $\dfrac{27}{3} = 9$
$5 = 5$	This means the answer is correct.

Quick Exercises 1 Solve each equation.

1) $1 + \dfrac{x}{2} = -2$

2) $\dfrac{x}{6} + (-2) = -7$

2-2. Solve the equation.

$$14 - \frac{x}{5} = 8$$

> SOLUTION

First, look at the equation. x represents the unknown value of the equation.

$14 - \frac{x}{5} = 8$	Original equation
$(-14) + 14 - \frac{x}{5} = 8 + (-14)$	Add (-14) to both sides.
$-\frac{x}{5} = -6$	Simplify. $(-14) + 14 = 0$, $8 + (-14) = -6$
$(-5) \times (-\frac{x}{5}) = (-6) \times (-5)$	Multiply each side by -5.
$x = 30$	Simplify. $(-5) \times (-\frac{x}{5}) = x$. $(-6) \times (-5) = 30$

So, the value of x is 30.

Check the solution.

$14 - \frac{x}{5} = 8$	Original equation
$14 - \frac{30}{5} = 8$	Substitute x with 30.
$14 - 6 = 8$	Simplify. $\frac{30}{5} = 6$
$8 = 8$	The answer is correct.

Quick Exercises 2 Solve each equation.

1) $-1 - \frac{x}{2} = -4$

2) $\frac{x}{2} + (-2) = 8$

Exercises 1   Solve each equation.

1) $-2 + x = 2$

2) $x + 5 = -6$

3) $-x + 1 = -9$

4) $-8 + y = 17$

Exercises 2   Solve each equation.

1) $x + 4 = -9$

2) $-8 + y = -9$

3) $\$34.45 + \dfrac{x}{6} = \$72.95$

4) $\dfrac{x}{2} + \$3.53 = \$16.64$

5) $\dfrac{y}{8} + 4 = -2$

6) $-10 + (y \div 2) = 15$

7) $-x + 10 = 11$

8) $-8 + \dfrac{y}{7} = 10$

9) $-4 + 5(x \div 6) = 6$

10) $2(x \div 5) + 11 = 13$

Exercises 3   Solve each expression.

1) If $x = -1$ and $y = -2$, find the value of $(2 - x)(2y - 3)$.

2) If $x = 1$ and $y = -2$, find the value of $(-2 + x)(2y + 3)$.

3) If $x = 3$ and $y = -3$, find the value of $(2 - x)(-1 - 2y)$.

4) If $x = -2$ and $y = 3$, find the value of $(-x - 1)(-3 - 2y)$.

5) If $x + \dfrac{2}{3} = -1$, find the value of $x + 1$.

6) If $-2 - x = \dfrac{1}{3}$, find the value of $2 + x$.

7) If $-(y + \dfrac{1}{5}) = 2$, find the value of $\dfrac{1}{5} + y$.

8) If $y - 1 = -\dfrac{2}{3}$, find the value of $y + \dfrac{4}{6}$.

Exercises 4   Solve each equation.

1) $0.5x + 1.2 = -6.8$

2) $-2x + 9 = 2 + 1$

3) $6 + 8 + \dfrac{1}{2} = 10 + x + 2$

4) $\dfrac{1}{4} + x = 2 + 4$

5) $2 + \dfrac{y}{2} = 4 + 12$

6) $\dfrac{x}{8} + 5 = 3 + 6$

7) $5 + \dfrac{x}{5} + 1 = 10 + 8$

8) $4 + \dfrac{y}{2} = 2 + \dfrac{y}{3}$

9) $12 + 4 = 6 + \dfrac{2y}{5}$

10) $\dfrac{3x}{8} + 10 = 2 + 6$

11) $2 + \dfrac{4x}{5} + 7 = 3 + 5$

12) $2 + \dfrac{5y}{4} = 1 + \dfrac{3y}{4}$

Exercises 5   Solve each expression.

1) If $x + 2 = \dfrac{1}{3}$ and $y = 3$, find the value of $(2 + x)(2y - 3)$.

2) If $x = \dfrac{2}{3} - 1$ and $y = 1 - \dfrac{1}{3}$, find the value of $(x + 1)(y + \dfrac{1}{3})$.

3) If $2y = 2 - \dfrac{4}{5}$ and $-(x - 2) = \dfrac{2}{4}$, find the value of $(\dfrac{2}{5} + y)(2 - x)$.

4) If $x = -2$ and $y = -4$, find the value of $(-2 + x)(y + 3)$.

Exercises 6   Solve each equation.

1) $-3 - x = -10$

2) $x - 8 = -20$

3) $y - 2 = 2$

4) $24 - y = -24$

5) $8 - \dfrac{y}{2} = 7$

6) $\dfrac{y}{5} - 1 = 9$

7) $\dfrac{x}{4} - 9 = 17$

8) $8 - \dfrac{x}{8} = -2$

9) $-2 - \dfrac{y}{2} = 5$

10) $\dfrac{y}{3} - 1 = -11$

11) $-\dfrac{x}{4} - 3 = 12$

12) $-5 - \dfrac{x}{2} = -2$

Exercises 7   Solve each equation.

1)   If $2x = -\dfrac{1}{2}$, find the value of $3 - x$.

2)   If $-x = \dfrac{2}{5}$, find the value of $x + 1$.

3)   If $9y = -2$, find the value of $\dfrac{1}{9} - y$.

4)   If $y - \dfrac{2}{3} = 3$, find the value of $2(y - \dfrac{2}{3})$.

5)   If $(\dfrac{1}{9} - y)^2 = 4$ and $x = \dfrac{1}{2}$, find the value of $(\dfrac{1}{9} - y)(x + 1)$.

6)   If $y = \dfrac{1}{7}$ and $x = \dfrac{1}{3}$, find the value of $(y - 1)(2 + x)$.

Exercises 8   Find the value of the variable in each equation.

1) $-10.8 + 2(5 - 2x) = 3.2$

2) $9(2x + 3) + 1 = -2$

3) $-2(8 + 0.5x) = -10$

4) $3 - (-2) - 3x = -4$

5) $-2 + 4(1 - x) = 10$

6) $(2x + 2) + 3 = -7$

7) $-(3 + 0.5x) = -2$

8) $-5 - 3x = 19$

Exercises 9   Solve each expression.

1)   If $x \div 2 = 1$, find the value of $3 - x$.

2)   If $2x \div 5 = 1$, find the value of $x - 1$.

3)   If $3y = 2$, find the value of $(1 \div 9) - y$.

4)   If $9y = 3$, find the value of $y - (2 \div 3)$.

5)   If $9y = 2$ and $x = \frac{1}{4}$, find the value of $(\frac{7}{9} - y)(x + 2)$.

6)   If $\frac{1}{9}y = 1$ and $x = \frac{1}{3}$, find the value of $(9x^2 + xy - 1)$.

7)   If $2x^2 = 32$ and $2y = -\frac{1}{4}$, find the value of $x^2y - 1$.

8)   If $2(x - 1) = -4$, find the value of $(x - 1)^2$.

* Solving Problems

Exercises 10   Solve each problem using the given information.

* For Exercises **1-2**. A candy factory makes 430 chocolate bars every 40 minutes.
1) What is the expression that describes how long it takes to make 1290 chocolate bars? Use x to represent the amount of time.

2) What is the value of x?

Exercises 11   Solve each problem using the given information.

* Grace has to go to school by 8:30 AM. It takes her 5 minutes to get dressed, 12 minutes to eat breakfast, 3 minutes to brush her teeth, and 15 minutes to walk to school.
1) Find the equation that shows how much time it takes for Grace to get to school on time.

2) What time does she have to get up in order to make it to school on time?

3) What if Grace was running late and got to school at 8:45? If she had stuck to her schedule, what time did she wake up?

4) If it took Grace 15 minutes more to do everything, what time does she have to get up now?

SELF-TEST

1. Which of the following equations represents the equation "the sum of −3 times x and 4 is 9"?

   A. $-3 \div 4x = 9$ B. $-3 - 4x = 9$
   C. $-3 + 9 = 4x$ D. $-3x + 4 = 9$

2. What is the value of x?
$$-24 + 3x = 15$$

   A. −13   B. 13
   C. 39   D. −39

3. Matt receives his salary from his company every two weeks. Currently he has $453.71 in his bank account. When he checked his bank account 3 months later, it shows that he now has $3261.54. Write an equation that describes how much money he received his salary from his company every two weeks. Use x to represent how much money he has.

   A. $453.71 + 2x = 3261.54$ B. $453.71 + 3(2x) = 3261.54$
   C. $453.71 + 3x = 3261.54$ D. $3261.54 + 3(2x) = 54453.71$

* Use the information for Exercise **4-5**. Oliver has $150.30 in his account. He starts a part-time job during the summer break at a restaurant and will receive $50.45 every week.

4. Write an equation that describes how many weeks he worked if he has $775.70. Use x to represent the number of weeks.

   A. $50.45 + 150.30x = 755.70$ B. $755.70 + 50.45x = 150.30$
   C. $150.30 + 50.45x = 755.70$ D. $150.30 - 50.45x = 755.70$

5. What is the value of x?

   A. 12   B. 8
   C. 4   D. 10

* Use the information for Exercises **6-7**.
   The fastest land animal in the world is the cheetah, An adult cheetah is capable of running up to 70 miles per hour, while a young cheetah is capable of running 45 miles per hour.

6. What is the expression used to find the total speed of several young and adult cheetahs? Use x to represent the speed of the young cheetahs and y for the number of the adult cheetahs.

A. $\dfrac{70\ miles}{hour}x + \dfrac{45\ miles}{hour}y$ B. $\dfrac{70\ miles}{hour}x - \dfrac{45\ miles}{hour}y$

C. $\dfrac{70\ miles}{hour}x \times \dfrac{45\ miles}{hour}y$ D. $\dfrac{70\ miles}{hour}x \div \dfrac{45\ miles}{hour}y$

7. If $x = 20$ minutes and $y = 10$ minutes, what is the value of the expression?

A. 15.8   B. 95
C. 185   D. 31

* Use the information for Exercises **8-9**. An ice cream factory is making ice cream sandwiches. They made 180 vanilla ice cream sandwiches, 80 strawberry ice cream sandwiches, and 270 chocolate ice cream sandwiches in 30 minutes.

8. Given that they produce the same amount of ice cream sandwiches at the same rate, what is the equation that shows how long it take for the factory to produce 240 strawberry ice cream sandwiches? Use x to represent the amount of time.

A. $80 \times 30x = 240$ B. $80 \div 30x = 240$
C. $80x \div 30 = 240$ D. $80 + 30x = 240$

9. How long does it take for the factory to produce 240 strawberry ice cream sandwiches?

A. 1h   B. 1h 30m
C. 2h   D. 1h 50m

10. Which of the following represents the expression "the difference of −2 times x and 5 is −12"?

A. $-2 \div 5x = -12$ B. $-2x - 5 = -12$
C. $-2 + 5x = -12$ D. $-2 - 12 = 5x$

11. What is the value of x for the equation below?
$$8x - 8 = -48$$

A. 5   B. 10
C. −5   D. −10

12. If $x = 2$ and $y = 5$, what is the value of the expression below?

$$2xy - (4y \div 5)$$

    A. 16   **B.** $16\frac{1}{5}$

    C. $18\frac{1}{5}$ **D.** 18

13. If $x = 2$ and $y = 0.5$, what is the value of the expression below?

$$(5.54 - 3y)(x + 1)$$

    A. −0.69   **B.** 12.12

    C. 1.04   **D.** 3.12

14. If $x = \frac{1}{3}$ and $y = 3$, what is the value of the expression below?

$$(4 - 3x)(y - 1)$$

    A. 2   **B.** 4

    C. 6   **D.** 8

15. What is the value of x for the equation below?

$$\tfrac{1}{2}x - 2 = -\tfrac{6}{9}$$

    A. $1\frac{4}{9}$ **B.** $1\frac{2}{3}$

    C. $2\frac{2}{3}$ **D.** 2

16. Which of the following expressions is equivalent to $2(x \div 3) + 4$?

    A. $2(\frac{1}{3}x + 2\frac{1}{3})$ **B.** $2(\frac{1}{3}x + 2)$

    C. $2(\frac{1}{3}x)$ **D.** $2(\frac{x}{3} + 4)$

17. On Monday, it rained $\frac{1}{4}$ inches. On Tuesday, it rained 2 inches. In total, it rained $8\frac{3}{4}$ inches for five consecutive days. Determine how many inches it rained for the next three days.

    A. $8\frac{1}{2}$ **B.** $8\frac{1}{4}$

    C. $6\frac{1}{4}$ **D.** $6\frac{1}{2}$

* For Exercises **18-19**. A cell phone repair company meets a quota of finishing 504 orders per day. 2/3 of the orders are smartphones while 1/3 of the orders are laptops.

18. What is the equation that shows the number of laptops fixed in 30 days? Use x to represent the number of laptops.

   A. $(504) \times (\frac{1}{3}) = x \times 30$ **B.** $(504) \times (\frac{1}{3}) = x + 30$

   C. $(504) \times (\frac{1}{3}) = x \div 30$ **D.** $(504) \times (\frac{1}{3}) = x - 30$

19. How many laptops are fixed in 30 days?

   A. 5,400   **B.** 15,120
   C. 210   **D.** 5,040

20. Lisa has \$32.00 in her account. She earns \$20.50 by babysitting for two hours. If she earns the same amount of money every hour, what is the equation that expresses how many hours Lisa has to babysit for her to have \$339.50 in her account? Use x to represent the number of hours.

   A. $20.50x + 2 = 339.50 - 32.00$ **B.** $20.50x \div 2 = 339.50 + 32.00$
   C. $20.50x \times 2 = 339.50 - 32.00$ **D.** $20.50x \div 2 = 339.50 - 32.00$

21. How many hours does Lisa have to babysit for her to have \$339.50 in her account?

   A. 23   **B.** 30
   C. 32   **D.** 19

* For Exercises **22-23**. There are 49 pieces of peppermint candy in the bowl. Kelly eats 2 pieces of candy and gave some candy to her friend. By the end of the day, there are 23 pieces left.

22. Which of the following equations correctly expresses the problem?

   A. $49 - 2 - x = 23$ **B.** $23 + x = 49 - 5$
   C. $49 - 2 + 23 = x$ **D.** $23 - 2 = x + 49$

23. How many pieces of candy does Kelly give to her friend?

   A. 22   **B.** 24
   C. 26   **D.** 28

2.   Solving Equation with One Variable: Multiplication and Division

2–3. Solve the equation.

$$x \div 11 = 1 \div 5 \ \text{ or } \ \frac{1}{11}x = \frac{1}{5}$$

SOLUTION

$\frac{1}{11}x = \frac{1}{5}$	Original equation
$11 \times \frac{x}{11} = \frac{1}{5} \times 11$	Multiply each side by 11.
$x = \frac{11}{5}$	Simplify. $11 \times \frac{x}{11} = x.$

So, the solution of $\frac{1}{11}x = \frac{1}{5}$ is $\frac{11}{5}$.

You can check the solution of $x = \frac{11}{5}$ by substituting the value in the original equation.

$\frac{1}{11}x = \frac{1}{5}$	Original equation
$\left(\frac{1}{11}\right)\left(\frac{11}{5}\right) = \frac{1}{5}$	Substitute x with $\left(\frac{11}{5}\right)$.
$\frac{1}{5} = \frac{1}{5}$	This means the value is correct.

Quick Exercises 3   Solve each expression.

1) $1.2 \div x = 0.3$ **2)** $(x \div 3) \times (-3) = -8$

2–4. Solve the equation.

$$\frac{1}{6} \div \frac{x}{2} = 4$$

SOLUTION

You can solve the problem in two ways (a) and (b).

a)

$\frac{1}{6} \div \frac{x}{2} = 4$

i) Flip the numerator and denominator (reciprocals).

ii) Change the operation.
iii) Then multiply.

$\frac{1}{6} \times \frac{2}{x} = 4$

$\frac{2}{6x} = 4$ Find the reciprocal of the divisor and multiply.

$$6x \times \frac{2}{6x} = 4 \times 6x \qquad \text{Multiply each side by } 6x.$$

$$2 = 24x \qquad \text{Simplify. } 6x \times \frac{2}{6x} = 2$$

$$\frac{1}{24} \times 2 = 24x \times \frac{1}{24} \qquad \text{Multiply each side by } \frac{1}{24}.$$

$$\frac{2}{24} = x \qquad \text{Simplify.}$$

$$\frac{1}{12} = x \qquad \text{Simplify.}$$

b) When solving an equation like $\frac{1}{6} \div \frac{x}{2} = 4$, you can manipulate it to be $\frac{1}{6} = \frac{x}{2} \times 4$.

$$\frac{1}{6} \div \frac{x}{2} = 4 \ \text{ or } \ \frac{1}{6} = \frac{x}{2} \times 4 \qquad \text{Rewrite the equation.}$$

$$\frac{1}{6} = 2x \qquad \text{Divide 2 and 4 with the GCF of 2.}$$

$$\frac{1}{2} \times \frac{1}{6} = 2x \times \frac{1}{2} \qquad \text{Multiply each side by } \frac{1}{2}.$$

$$\frac{1}{12} = x \qquad \text{Simplify.}$$

So, the solution of $\frac{1}{6} \div \frac{x}{2} = 4$ is $x = 1/12$.

2-5. Equivalent and Reciprocals.

SOLUTION

a) You can solve the problem both ways and the answer will be the same.
* Remember that you can rewrite the equations.

dividend ÷ divisor = quotient
⇕   ⇕
dividend = divisor × quotient

b) Reciprocals: Two numbers are reciprocals if their product is 1. For example:

$$\frac{1}{2} \times \frac{2}{1} = 1 \qquad \frac{5}{7} \times \frac{7}{5} = 1 \qquad \frac{9}{10} \times \frac{10}{9} = 1$$

Quick Exercises 4   Solve each expression.

1) $3 \div \frac{x}{2} = 4$

2) $2 \div \frac{x}{7} = 1\frac{1}{6}$

2–6. If $x = -2$, what is the value?

$$5 + x \div 5$$

SOLUTION

$5 + x \div 5$	Original expression
$5 + \dfrac{x}{5}$	Rewrite the expression.
$5 + \dfrac{-2}{5}$	Substitute x with -2.
$4\dfrac{5}{5} - \dfrac{2}{5}$	Rewrite the expression. $5 = \dfrac{25}{5} = 4\dfrac{5}{5}$
$4\dfrac{3}{5}$	Simplify. $4\dfrac{5-2}{5} = 4\dfrac{3}{5}$

So, the value of the expression is $4\dfrac{3}{5}$.

Exercises 12   Solve each equation.

1) $0.5x = -12$

2) $1.2y = -7.2$

3) $-2y = 0.2$

4) $-0.1x = 1$

5) $3x = 9.6$

6) $2.5y = 3$

7) $7x = -14.7$

8) $0.5x + 3 = 7$

9) $2y - 14 = 2$

10) $8x + 1 = -7$

11) $7 - 3x = 13$

12) $2(20 + 6y) = -8$

Exercises 13   Find the value using the given information.

1) If $y = -\dfrac{4}{5}$, find the value of $\dfrac{5}{8}y$.

2) If $x = \dfrac{2}{3}$, find the value of $\dfrac{6}{8}x$.

3) If $y = -0.4$, find the value of $(-2y)$.

4) If $y = -3.5$, find the value of $2y$.

5) If $y = 2$, find the value of $4y - 2$.

6) If $y = -4$, find the value of $16 - 2y$.

Exercises 14   Solve each equation.

1) $4x = -24$

2) $3y = -18$

3) $-5y = 2$

4) $-10x = 110$

5) $\dfrac{x}{4} = 96$

6) $(\dfrac{y}{6})(14) = 84$

7) $(\dfrac{y}{10})(3) = -21$

8) $\dfrac{y}{3} = 7$

9) $\dfrac{1}{5}y = \dfrac{1}{10}$

10) $(\dfrac{x}{8})(5) = -1\dfrac{1}{4}$

11) $\dfrac{1}{9}y - 5 = 2$

12) $5(1 + \dfrac{y}{5}) = 1$

Exercises 15   Find the value using the given information.

1) If $2y = 10$, find the value of $2(y - 1)$.

2) If $3y = -1$, find the value of $0.4 - 3y$.

_____

3) If $2(y - 1) = -8$, find the value of $(-2y + 2) \times (-4)$.

4) If $\frac{2}{3}y = -2$, find the value of $(\$8.50)y + \$30.50$.

_____

5) If $\frac{3}{4}x = -2$, find the value of $2(3x + 6)$.

6) If $-(2y - 3) = 5$, find the value of $2(3 - 2y)$.

_____

Exercises 16   Solve each equation.

1) $84 \div y = 14$

2) $17 \div x = 1$

_____

3) $2x \div 62 = 8.4$

4) $3.2 \div 4y = 8$

_____

5) $44 \div 2.2x = -4$

6) $0.5x \div 1.5 = -3$

_____

7) $8y \div 12 = 4$

8) $3 \div 5y = 2$

_____

9) $12 \div 2x = 4$

10) $-8.64 \div 4.8y = 3$

_____

11) $(-8) \div (-2x) = -2$

12) $(-8) \div 4y = -4$

_____

Exercises 17   Find the value of the expression using the given information.

1)   If $y = 10$, find the value of $0.7y - 2$.

2)   If $y = -0.5$, find the value of $0.4 - 3y$.

3)   If $x = -8$, find the value of $(-1.6x)$.

4)   If $y = -3$, find the value of $(\$8.50)y + \$1.50$.

5)   If $x = 2$, find the value of $2(2x - 1)$.

6)   If $y = 5$, find the value of $2(3 - 2y)$.

Exercises 18   Solve each equation.

1) $-16 \div y = 8$

2) $15 \div x = -10$

3) $x \div (-15) = 4$

4) $52 \div y = (-2)$

5) $3 \div \dfrac{x}{8} = -8$

6) $\dfrac{x}{8} \div 2 = -3$

7) $\dfrac{y}{2} \div 5 = 5$

8) $4 \div \dfrac{x}{2} = 4$

9) $9 \div \dfrac{y}{2} = 4$

10) $5 \div \dfrac{y}{7} = 21$

11) $\dfrac{2}{3} \div \dfrac{x}{8} = -8$

12) $\dfrac{1}{8} \div \dfrac{y}{2} = -4$

Exercises 19   Find each value.

1) If $x = 2$, find the value of $4 \times \dfrac{x}{8} + 2$.

2) If $y = 5$, find the value of $3 \times \dfrac{x}{5}$.

3) If $y = 6$, find the value of $2 - \dfrac{y}{4}$.

4) If $x = 1$, find the value of $1 - \dfrac{x}{3}$.

5) If $x = 5$, find the value of $2 \times \dfrac{x}{3}$.

6) If $y = 4$, find the value of $\dfrac{y}{3} \div 2$.

7) If $x = 2$, find the value of $\dfrac{1}{2} \div \dfrac{x}{8}$.

8) If $y = 4$, find the value of $\dfrac{y}{9} \div \dfrac{8}{18}$.

9) If $2(y \div 3) = 5$, find the value of $\dfrac{1}{3}y - \dfrac{1}{2}$.

10) If $2 \div x = 6$, find the value of $\dfrac{1}{3} + \dfrac{2}{x}$.

Exercises 20   Find the value using the given information.

1) If $y \div 3 = -0.2$, find the value of $0.7(y \div 3)$.

2) If $\dfrac{4}{10} - 3y = -0.5$, find the value of $0.4 - 3y$.

3) If $2(5 \div x) = -8$, find the value of $(-1.6x)$.

4) If $\dfrac{x}{3} = -2$, find the value of $(\$8.50)x + (-\$20.50)$.

5) If $\dfrac{x - 3}{2} = 2$, find the value of $2(x - 3)$.

6) If $\dfrac{4 + x}{4} = 1$, find the value of $1 + (x \div 4)$.

* Solving Problems

Exercises 21   Solve each problem using the given information.

Alai and four of his friends ate at a restaurant. The bill for the meal was $62.80. Including tips, they paid $75.36.
1) What is the percentage of the bill that represents tips?

2) If Alai paid 25% of the total bill, write the equation that describes how much money he paid.

Exercises 22   Solve each problem using the given information.

Bob has 103 toy cars that he would like to keep in several boxes. He would like to put 8 toy cars in each box.
1)   Which equation could be used to find that how many boxes needed in order to keep 103 toy cars?

2)   How many boxes will he use?

Exercises 23   Solve each problem using the given information.

1)   The school is planning a field trip. There are 410 students in total and 35 students in each bus, but two buses each have 5 less students than the other buses. What equation could be used to determine many buses will be needed to take every student?

2)   How many buses will be needed to take every student?

Exercises 24   Solve each problem using the given information.

Use the information for Exercises 1-3. A miser is organizing his coins. He has quarters and dimes in several bags. Each bag has 7 less quarters than dimes.
1)   What expression could be used to solve how much money is in each bag? Use x to represent the number of quarters and y to represent the number of dimes. Show your work.

2) What are the values of x and y if there are 12 bags of coins?

3) How much money are there in 12 bags?

Exercises 25   Solve each problem using the given information.

Use the information for Exercises **1-2**. At a fundraiser, 9 people donated some money. Eight of them donated the same amount of money except for one person who donated $25.00 more than the others.

1) If the money they donated adds up to $340.00, how did you set up the equation showing how much each person donated? Use x to represent the amount of money.

2) How much money did each person donate except one person?

SELF-TEST

1. Which of the following represents the expression "the product of 4 times x and 2 is 54"?

    A. $4 \div 2x = 54$ **B.** $4 \times 54 = 2x$
    C. $4 + 2x = 54$ **D.** $4x \times 2 = 54$

2. Which of the following represents the expression "the product of N and 1 is 14"?

    A. $N + 1 = 14$ **B.** $N \div 1 = 14$
    C. $N = 14$ **D.** $N - 1 = 14$

3. What is the value of x for the equation below?
$$-\frac{x}{4} = -20$$

    A. 10   **B.** -10
    C. 80   **D.** -80

4. If a circumference of a circle is $1\frac{1}{4}$ cm, write an expression showing how to find the length of the radius. Use r for the length of the radius.

A. $1\frac{1}{4} = 2\pi r$ B. $1\frac{1}{4} \times r = 2\pi$

C. $1\frac{1}{4} - r = 2\pi$ D. $1\frac{1}{4} + r = 2\pi$

5. Marcus is dividing 54 pieces of M&Ms candy between his two brothers. His younger brother will get half as much as his older brother. Which expression could be used to show how many pieces each brother will get? Use x to represent the number of M&Ms.

A. $\frac{1}{2}x - x = 54$ B. $\frac{1}{2}x + x = 54$

C. $\frac{1}{2}x \times x = 54$ D. $\frac{1}{2}x \div x = 54$

6. The bakery is baking two kinds of loaves of banana bread. The first loaf has 6 bananas and the second loaf has twice as many as the first loaf. What is the equation to showing how many bananas are in the first and the second loaves? Use x to represent the number of bananas.

A. $6x + \frac{1}{2}x$ B. $6(x + 2x)$

C. $\frac{1}{2}x + 6x$ D. $2(6x) + 6x$

* Use the information for Exercises 7-9. A miser is organizing his coins. He is putting quarters and dimes in several bags. Each bag has 5 more quarters than dimes.

7. What equation could be used to solve how much money is in each bag? Use x to represent the number of quarters and y to represent the number of dimes.

A. $5 + 0.25x + 0.1y$ B. $0.25(5 + x) + 0.1y$

C. $0.25x + (5 + 0.1y)$ D. $5(0.25x) + 0.1y$

8. What are the values of x and y if there are 9 bags?

A. $x = 3.50, y = 0.9$ B. $x = 3.00, y = 0.9$

C. $x = 2.50, y = 0.9$ D. $x = 2.00, y = 0.9$

9. How much money is there in 9 bags?

        A. $4.40   **B.** $3.15
        C. $2.65   **D.** $1.40

* Use the information for Exercises **10-11**. At a fundraiser, 9 people each donated the same amount of money while another person donated $12.00 less than the other people.

10. If the money they donated adds up to $250.00, how did you set up the equation to show how much money each person donated? Use x to represent the money that was donated.

        A. $9x + (x - 12) = 250$ **B.** $9x + (x + 12) = 250$
        C. $9x \times (x - 12) = 250$ **D.** $9x - (x + 12) = 250$

11. How much money did all 9 people donate?

        A. $235.00   **B.** $245.00
        C. $225.00   **D.** $255.00

12. A table is set out for a buffet. On the table are plates of crab. Each plate has exactly 5 pieces of crab except for a plate that has twice as many pieces. There are 225 pieces of crab in total. What is the equation used to find how many plates of crab are set out?

        A. $5x = 225$ **B.** $225 = x \div 5$
        C. $225x = 5$ **D.** $225 = 5 \div x$

13. Which of the following represents the expression "the quotient of 2 times x and 2 is 6"?

        A. $2 \times 2x = 62x$ **B.** $2 - 2x = 6$
        C. $2 + 2x = 6$ **D.** $2x \div 2 = 6$

14. Which of the following represents the expression "the quotient of 81 and 3 is y"?

        A. $3 \div 81 = y$ **B.** $y = 81 \times 3$
        C. $3 + y = 81$ **D.** $y = 81 \div 3$

15. David and his four friends went a restaurant and equally shared the bill of $79.85 for their meal. Write an expression describing how much money each person gave. Use

y to represent the amount of money.

A. $y \div 79.85 = 5$
C. $5y = 79.85$

B. $y \div 5 = 79.85$
D. $y = 79.85 \times 5$

16. What is the value of x for the equation below?
$$5 \div (7x) = \frac{5}{7}$$

A. 1
C. 3

B. 2
D. 4

17. What is the value of x for the equation below?
$$x \div 8 = 4\frac{3}{8}$$

A. 64
C. 35

B. 12
D. 280

* Use the information for Exercises **18-19**. The school is planning a field trip. There are 288 students in total and 35 students in each school bus, but two school buses each have 4 more students than the other buses.

18. Which equation could be used to show how many buses are needed to take every student?

A. $288 = 35x + 2$
C. $288 = 35x + 4$

B. $288 = 35x + 2(4)$
D. $288 = 35x - 2(4)$

19. How many buses will be needed to take every student?

A. 6
C. 8

B. 7
D. 9

20. If the quotient of an equation is 12, what is the value of the divisor given that the dividend is 8?

A. $\frac{1}{2}$

B. $\frac{1}{3}$

C. $\frac{1}{4}$

D. $\frac{2}{3}$

21. If $k = 8$, what is the value of $\frac{9}{15} - (k + 6)$?

   A. $-1\frac{2}{5}$ B. $1\frac{2}{5}$

   C. $-\frac{9}{15}$ D. $\frac{9}{15}$

22. If $z = \frac{2}{3}$, what is the value of $(3 - z) + 6$?

   A. 6   B. $8\frac{1}{3}$

   C. $\frac{1}{3}$ D. $8\frac{2}{3}$

23. Given the expression "the quotient of $\frac{2}{3}$ and x increased by 2", what is the value if $x = 2$?

   A. $1\frac{1}{3}$ B. $\frac{5}{12}$

   C. $2\frac{1}{3}$ D. $\frac{1}{2}$

24. Given the expression "the product of 4 and K decreased by $\frac{1}{2}$", what is the value if $K = \frac{1}{4}$?

   A. 2   B. $\frac{1}{2}$

   C. $4\frac{1}{2}$ D. 8

25. A container in the shape of a square box can hold $4\frac{1}{6}$ pounds of powder. What is the expression that describes how many containers required if 25 pounds of powder are needed? Use x to represent the number of containers.

   A. $25x - 6 = 25$ B. $25x \times 6 = 25$
   C. $25x \div 6 = 25$ D. $25x \div 25 = 6$

26. Sam paid 20% of the total bill at a restaurant. He paid $7.50. What is the expression

that shows how much the bill is? Use x to represent the amount of money paid.

A. $(0.2) + (7.50) = x$
C. $0.2 + x = 7.50$

B. $(0.2)(7.50) = x$
D. $7.50 - x = 0.2$

27. Given the expression "the sum of $\frac{5}{7}$ and N decreased by 1", what is the value if $x = 2$?

A. $\frac{5}{35}$

B. $\frac{5}{7}$

C. 10

D. $1\frac{5}{7}$

28. Given the expression "the difference between $\frac{x}{4}$ and 3 divided by 2", what is the value if $x = 15$?

A. $-\frac{3}{8}$

B. $\frac{3}{8}$

C. $3\frac{3}{8}$

D. $1\frac{2}{4}$

29. If $z = 7$, what is the value of $\frac{5}{7}z - 6$?

A. 1

B. -1

C. $-\frac{6}{7}$

D. 32

30. If $x = 3$, what is the value of $(14 - 2x) \times \frac{1}{8}$?

A. 1

B. 2

C. 3

D. 4

3. Two Variables in Addition, Subtraction, Multiplication, and Division Problems

2–7. If $x = 2$, what is the value of y in the equation below?

$$7 = 3x + \frac{y}{5}$$

SOLUTION

a) There are two variables in the given equation, but only x is given. So, you can replace x with 2 and then continue to solve the equation.

$7 = 3 \times 2 + \dfrac{y}{5}$ Substitute 2 for x.

$7 = 6 + \dfrac{y}{5}$ Simplify.

$(-6) + 7 = 6 + (-6) + \dfrac{y}{5}$ Add (-6) to both sides.

$1 = \dfrac{y}{5}$ Simplify.

If you would like to solve for y in the $1 = \dfrac{y}{5}$, first multiply 5 on both sides.

$5 \times 1 = \dfrac{y}{5} \times 5$ Multiply each side by 5.

$5 = y$ Simplify.

So, the value of y is 5.

Quick Exercises 5 Solve each expression.

1) If $y = -1$, $3 = x + \dfrac{y}{2}$ 2) If $x = 2$, $-1 = x - \dfrac{y}{3}$

Exercises 26   Find the value of the unknown variable in each equation.

1) If $y = 2$, $\dfrac{x}{2} + y = 4$ 2) If $x = 3$, $2x + \dfrac{y}{3} = 1$

3) If $y = 2$, $3 = \dfrac{y}{4} + x$ 4) If $x = 9$, $y + 5 = \dfrac{x}{3}$

5) If $x = 5$, $10 = \dfrac{x}{5} + \dfrac{y}{2}$ 6) If $y = 4$, $7 = 5x + \dfrac{y}{2}$

Exercises 27   Find the value of the unknown variable in each equation.

1) If $x = 2, \dfrac{y}{2} - 4x = 2$

2) If $y = 5, y - \dfrac{x}{5} = 2$

3) If $y = 9, 2y - \dfrac{x}{2} = 16$

4) If $x = 2, 5x - \dfrac{1}{2} = \dfrac{y}{2}$

5) If $x = 4, x\dfrac{1}{2} - 2y = 3$

6) If $y = 4, 5x - 3y = 3$

* Solving Problems

Exercises 28   Solve each problem using the given information.

* Use the information for Exercises **1-3**. Bill needs to buy school supplies before school begins. He buys notebooks that costs $0.75 and glue sticks that costs $0.55 each.

1) Write an expression that describes how much money he needs to pay. Use x to represent the number of notebooks and y to represent the number of glue sticks.

2) How many notebooks did Bill buy if he bought 5 glue sticks and paid $12.50 in total?

3) If Bill bought 4 times as many glue sticks than notebooks, how many notebooks did he buy if he paid $12.50?

4. Conversion of Measurements

2-8. Convert the measurements.

$$1.8 \text{ lb} = \underline{\hspace{3cm}} \text{ ounces}$$

SOLUTION

First, look at the given units of the problem and then set up a proportion of two ratios with the same units.

i) Set up a ratio using the given information.

   Let x be the unknown quantity as 1.8 lb = x ounces.

$$\frac{1.8 \text{ lb}}{x \text{ ounces}}$$

You should know 1 pound (lb) = 16 ounces (oz).

ii) Set up a proportion of two ratios with the same unit.

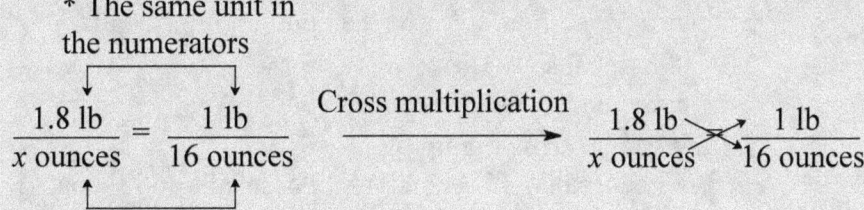

* The same unit in the numerators

$$\frac{1.8 \text{ lb}}{x \text{ ounces}} = \frac{1 \text{ lb}}{16 \text{ ounces}} \quad \xrightarrow{\text{Cross multiplication}} \quad \frac{1.8 \text{ lb}}{x \text{ ounces}} \times \frac{1 \text{ lb}}{16 \text{ ounces}}$$

* The same unit in the denominators

iii) Write a proportion.

   $(1.8 \text{ lb})(16 \text{ ounces}) = (x \text{ ounces})(1 \text{ lb})$

   The units can cancel out each other.

   $(1.8 \text{ lb})(16 \text{ ounces}) = (x \text{ ounces})(1 \text{ lb}) \implies (1.8)(16) = (x)(1)$

   The units can cancel out each other.

   $28.8 = x$ Multiply.

   So, 1.8 lb = __28.8__ ounces

2-9. Convert the measurements.

$$130 \text{ inches} = \underline{\hspace{3cm}} \text{ m}$$

SOLUTION

130 inches = _____ m

You should know the customary units of 1 inch (in.) = 0.0254 meters (m).

$$\frac{130 \text{ inches}}{x \text{ meters}} = \frac{1 \text{ inch}}{0.0254 \text{ meters}}$$ Set up a proportion of two ratios.

$(130 \text{ inches})(0.0254 \text{ m}) = (x \text{ m})(1 \text{ inch})$ Cross multiply.
$3.302 = x$ Cancel the units/ Multiply.
So, 130 inches equal 3.302 m.

2–10. An automatic machine can throw 48 baseballs per minute.
How many baseballs can the machine throw in 15 seconds?

SOLUTION

i) Look at the given information and then set up the ratio.

$$\frac{48 \text{ baseballs}}{1 \text{ minute}} \text{ or } \frac{48 \text{ baseballs}}{60 \text{ seconds}}$$

ii) Set up a proportion. Let x be the number of baseballs.

$$\frac{48 \text{ baseballs}}{60 \text{ seconds}} = \frac{x \text{ baseballs}}{15 \text{ seconds}}$$

iii) Rewrite and cross multiply.

$(48 \text{ baseballs})(15 \text{ seconds}) = (x \text{ baseballs})(60 \text{ seconds})$

Write as a proportion
$720 = (x)(60)$ Cancel the units/ multiply 48 and 15.
$12 = x$ Divide 60 from both sides.
So, the machine throws 12 baseballs in 15 seconds.

2–11. An automatic machine can throw 48 baseballs per minute.
How long will it take to throw 350 baseballs?

SOLUTION

The same way is used to solve the problem.

$$\frac{48 \text{ baseballs}}{1 \text{ minute}} = \frac{350 \text{ baseballs}}{x \text{ minutes}}$$ Set up a proportion of two ratios using the same units.

$(48 \text{ baseballs})(x \text{ minutes}) = (350 \text{ baseballs})(1 \text{ minute})$ Write as a proportion

$x \approx 7.29$ Cancel out the units each other/Divide 48 from both sides.

So, the machine throws 350 baseballs in about 7.29 minutes.

Quick Exercises 6 Solve each expression.

1) 132 Ibs = _____ Kg   **2)** 3.2 ft = _____ in.

2–12. Name of Units.

Lengths	Volume
1 yard (yd.) = 3 feet (ft) = 36 inches (in.)	1 kiloliter (kL) = 1,000 liters (L)
1 kilometer (km) = 1000 meter (m)	1 hectoliter (hL) = 100 liters (L)
1 meter (m) = 100 centimeters (cm) =	1 dekaliter (daL) = 10 liters (L)
1,000 millimeters (mm)	1 liter (L) = 10 deciliters (dL)
1 mile (mi.) = 1,760 yards (yd.) = 5,280	1 liter (L) = 100 centiliters (cL)
feet (ft)	1 liter (L) = 1,000 milliliters (mL)
Volume	**Mass (weights)**
1 pint = 2 cups (C)	1 kilogram (kg) = 1,000 grams (g)
1 quart = 2 pints (pt)	1 kilogram (kg) = 2.204 pounds
1 gallon = 4 quarts (qt)	1 pound (Ib) = 16 ounces (oz)
1 cup = 8 fluid ounces (fl oz)	

Exercises 29   Convert the measurements.

1)   1514 cm = _____ m

2) $\frac{1}{4}$ lb = _____ ounces (oz)

3) $\frac{1}{2}$ kg = _____ pounds (lb)

4)   412 pints = _____ quarts

5)   24 lb = _____ ounces (oz)

6)   254 pounds (lb) = _____ kg

7)   168 cm = _____ m

8)   15 feet (ft) = _____ in.

9)   1641 g = _____ kg

10)   0.35 feet (ft) = _____ in.

11)   12 quarts = _____ gallon

12)   18 fl oz = _____ pints

Exercises 30   A. Convert the measurements.

1)   7524 m = _____km

2)   68 yards (yd.) = _____ feet (ft)

3)   423 quarts = _____ pints

4) $2\frac{1}{4}$ yards (yd.) = _____ inches (in.)

5)   124 pints = _____cups

6)   76322 mL = _____ liters (L)

7)   540 gallons (gal) = _____ pints

8)   61.2 inches = _____ yards (yd.)

9)   13.5 cups = _____ fl oz

10)   3672 mL = _____ L

B. Convert the measurements.

1)   42 ft = _____yd.

2)   4 gal = _____ qt

3)   360 g = _____ oz

4)   30 pt = _____ gal

5)   240 lb = _____oz

6)   7.6 L = _____ mL

7)   12 gal = _____ pt

8)   100 yd= _____ ft

1. If $x = 3$, what is the value of $(4 \div 2) - N(x - 2) = 2(N \div 2)$?

    A. 0

    B. 1

    C. 2

    D. 3

2. Which of the following expression is equivalent to $(x - 2) - N(x - 2)$?

    A. $(1 \times N)(x - 2)$

    B. $N(x - 2)$

    C. $(1 + N)(x - 2)$

    D. $(1 - N)(x - 2)$

3. If $-k = 6$, what is the value of $\dfrac{4}{9} - (k + 7) = 5N \div 27$?

    A. 3

    B. -5

    C. -3

    D. -15

4. If $c = 6$, what is the value of $N - \dfrac{2}{3}(c) = 7$?

    A. 9

    B. 11

    C. 13

    D. 15

5. Lily went to the post office. If the rates of service are $0.41 for a regular mail and $3.51 for an express mail and she wants to send several letters of each, write an expression that describing the rates of service. Use x to represent the rate of regular mail and y for the rate of express mail.

    A. $0.41x - 3.51y$

    B. $0.41(x + y) + 3.51(x + y)$

    C. $0.41x + 3.51y$

    D. $0.41x \times 3.51y$

6. If $z = 15$, what is the value of $19 + k + \dfrac{5}{6}(z) = 83$?

    A. $1\dfrac{1}{3}$

    B. $\dfrac{2}{3}$

    C. $\dfrac{3}{4}$

    D. $\dfrac{1}{2}$

7. If $k = 2$, what is the value of $57 - (\frac{1}{2} \times k + 7) = N$?

    A. 47   **B.** 49
    C. 44   **D.** 45

8. Add the product of $\frac{2}{5}$ and K to the product 4 and N. Given that the sum is 20, write an expression that describes how to find the values of K and N.

    A. $(\frac{2}{5} + k) + (4 \times N) = 20$ **B.** $(\frac{2}{5} \times k) + (4 + N) = 20$

    C. $(\frac{2}{5} + k) + (4 + N) = 20$ **D.** $(\frac{2}{5} \times k) + (4 \times N) = 20$

9. Bill needs to buy school supplies before school begins. He bought some notebooks, some graph paper, and a glue stick. Given that the notebooks each cost $0.89, the graph paper costs $1.20, and the glue stick cost $.75, write an expression that describes how much money you need to pay. Use x to represent the number of notebooks and y to represent the number of graph paper.

    A. $1.20y + 0.75 + 0.89y$ **B.** $0.75(x + y) + 0.89(x + y) = 1.20$
    C. $1.20 + 0.75x - 0.89y$ **D.** $1.20 + 0.75x \times 0.89y$

10. Multiply the sum of 5 and N with the difference of $\frac{3}{4}$ and K. If the value is 14, which of the following is corresponds to the sentence?

    A. $(5 + N) - (\frac{3}{4} - K) = 14$ **B.** $(5 \times N) + (\frac{3}{4} - K) = 14$

    C. $(5 + N) \times (\frac{3}{4} - K) = 14$ **D.** $(5 \times N) \div (\frac{3}{4} - K) = 14$

11. Subtract the quotient of $3\frac{1}{2}$ and K with the product $\frac{1}{4}$ and N. If the value is 6 given $K = 2$, what is the value of N?

    A. $1\frac{1}{4}$ **B.** 13

    C. $2\frac{1}{2}$ **D.** -13

5.   Making an Equation: Table and Graph

2–13.   Use the table to make the graph. Explain the table and graph.

x	−6	−3	0	3
y	−6	−3	0	3

SOLUTION

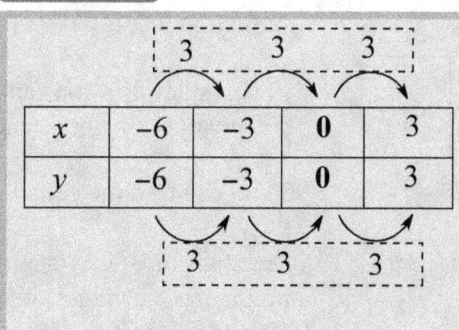

i) When $x = 0$, $y = 0$.
So, you can write the equation that is $y = ?\,x + 0$ or $y = ?\,x$.

ii) The values of x constantly increase by **3**.
The values of y constantly increase by **3**.

So the equation in the function table shows $3y = 3x$ or $y = x$
The values of x and y are equal to each other.

a)   The table gives the coordinates of x and y for $y = x$.

b)   Plot (x-coordinate, y-coordinate) on the grid.

c)   The coordinates given determines the relationship between x and y. Plots the points on the graph.

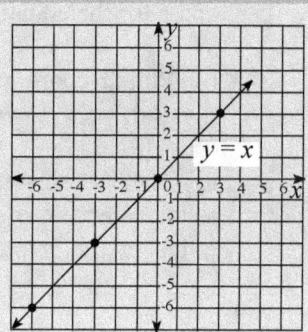

Quick Exercises 7   Use the table to make the graph. Explain the table and graph.

x	−1	0	1	2
y	−3	0	3	6

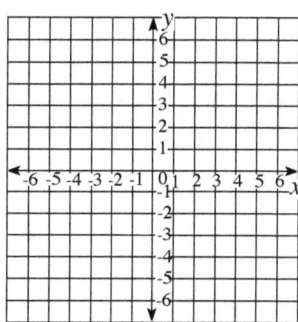

2-14. Use the table to make a graph. Explain the table and graph.

x	-2	-1	0	1
y	-5	-3	-1	1

SOLUTION

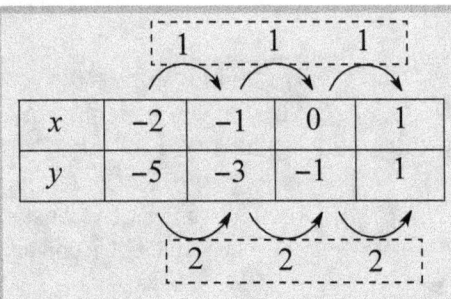

i) When $x = 0$, $y = -1$. So, you can write the equation that is $y = ? x - 1$.

ii) The values of x constantly increase by **1**. The values of y constantly increase by **2**.

So you can find the equation, which is $y = 2x - 1$.

The values of x and y change by the double amount and the values of y shift down by 1 from the point $(0, 0)$.

a) The table gives the coordinates of x and y for the equation $y = 2x - 1$.

b) Plot (x-coordinate, y-coordinate) on the grid.

c) The coordinates given determines the relationship between x and y. Plot the points on the graph.

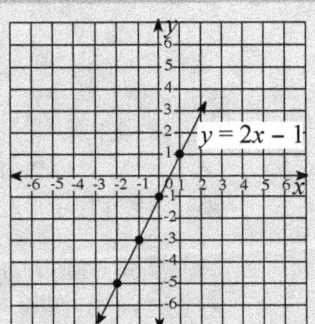

Quick Exercises 8   Use the table to make the graph. Explain the table and graph.

x	-1	0	1	2
y	-2	-1	0	1

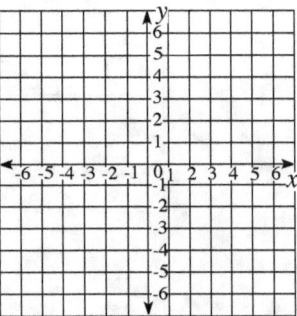

Exercises 31   Use the tables to make a graph and to find the equation. Explain the table and graph.

1)

x	-2	-1	0	1
y	-6	-3	0	3

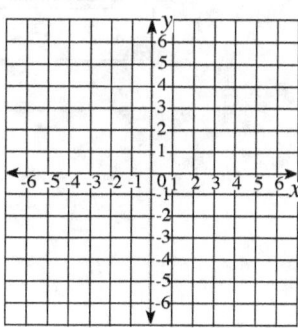

2)

x	-4	-3	-2	-1
y	4	3	2	1

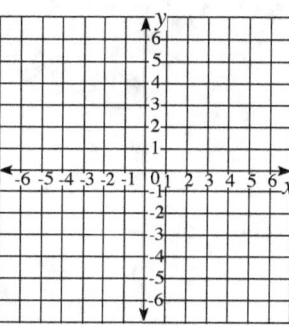

3)

x	-2	-1	0	1
y	6	3	0	-3

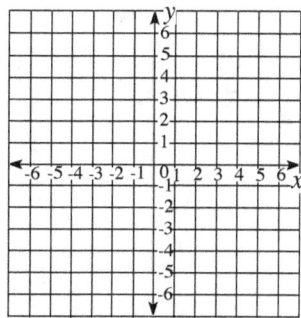

4)

x	-4	-2	0	2
y	-8	-4	0	4

Exercises 32   Use the tables to make a graph and to find the equation. Explain the
table and graph.

1)

x	−4	−2	0	2
y	−3	−1	1	3

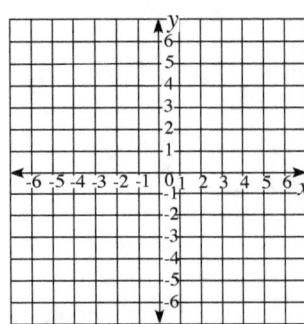

2)

x	−2	−1	0	1
y	3	2	1	0

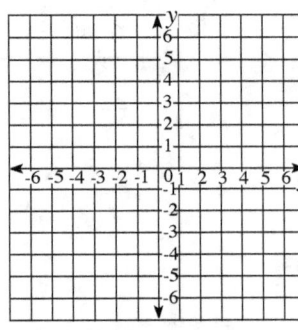

3)

x	−4	−2	0	2
y	−7	−3	1	5

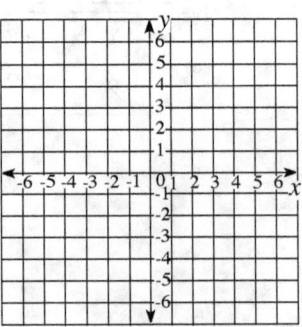

4)

x	−2	−1	0	1
y	−4	−1	2	5

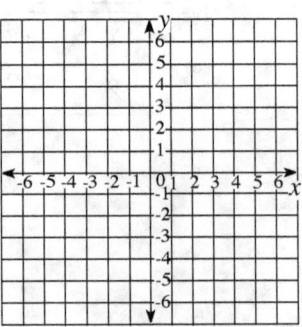

For Exercises **1-2**, use the graph below.

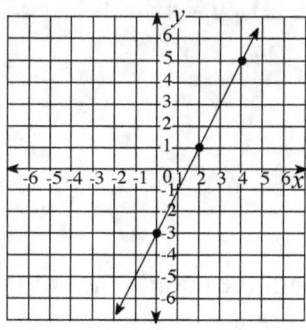

1. The graph shows the coordinates of (0, –3) and (2, 1). Which of the following coordinates are on the same line?

    A. (2, 2)   **B.** (2, 4)
    C. (3, 4)   **D.** (3, 3)

2. Which of the following represents the equation of the graph?

    A. $y = 2x + 1$ **B.** $y = 2x - 3$
    C. $y = 3x - 1$ **D.** $y = 3x - 2$

3. There is a linear function with the coordinates of (–2, 4), (0, 3), and (4, 1). What is the equation of the graph?

    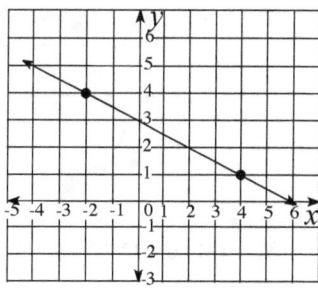

    A. $y = 2x + 2$ **B.** $y = -\frac{1}{2}x + 2$

    C. $y = -\frac{1}{2}x$ **D.** $y = 4x - 2$

4. There is a linear function with the coordinates of (–3, –1), (0, 2), and (3, 5). What is the equation of the graph?

A. $y = 2x + 2$ **B.** $y = x + 2$
C. $y = 2x$ **D.** $y = 4x - 2$

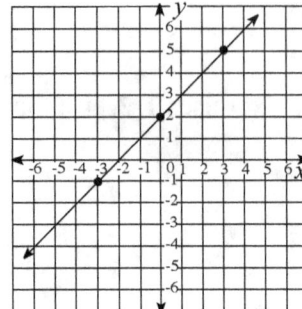

5. There is a linear function with the coordinates of $(-4, 2)$, $(-2, 0)$, $(0, -2)$, and $(3, -4)$. What is the equation of the graph?

A. $y = -2x + 2$ **B.** $y = x + 2$
C. $y = 2x$ **D.** $y = -x - 2$

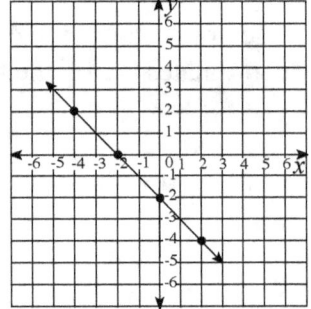

6. Which of the following represents the equation $y = 2x - 1$?

A.

x	0	2	4
y	−1	4	8

B.

x	−2	0	2
y	−3	−1	3

C.

x	2	4	6
y	3	7	11

D.

x	−4	0	1
y	−9	0	1

7. Which of the following represents the equation $4y - 3 = x$?

A.

x	0	2	4
y	−1	0	1

B.

x	−2	0	2
y	−2	−1	0

C.

x	5	9	13
y	2	3	4

D.

x	−4	0	1
y	−2	0	2

8. Which of the following represents the equation $y = 2x + 1$?

A.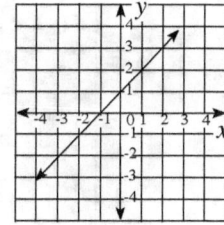

B.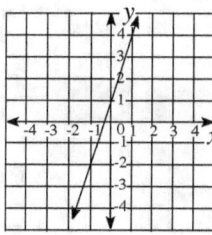

C.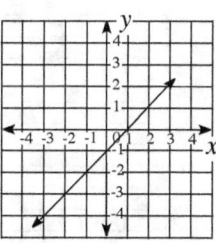

D.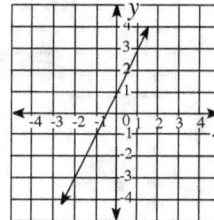

9. Which of the following represents the equation $y = 3x + 1$?

A.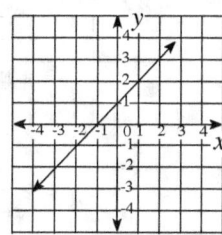

B.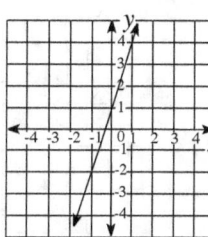

C.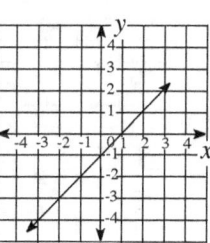

D.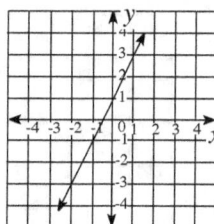

10. What is the equation of the graph below?

A. $y = 3$ B. $x = 3$
C. $y = -3$ D. $x = -3$

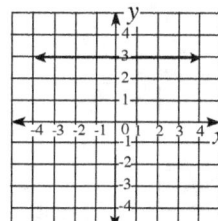

11. Which of the following graphs shows an increase in the values?

A.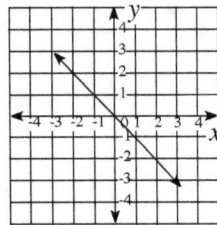

B.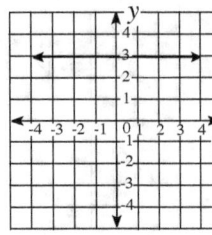

C.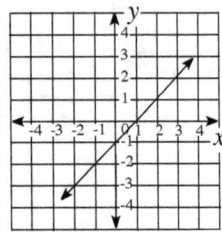

D.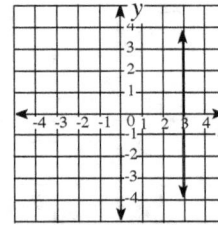

12. Which of the following graphs shows a decrease in the values?

A.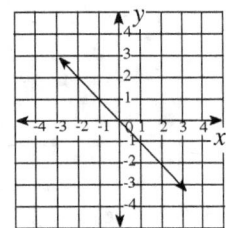

B.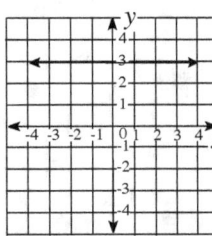

C.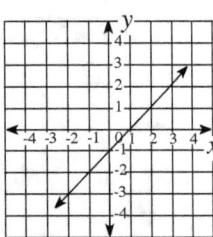

D.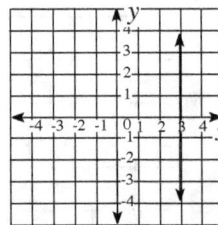

13. What is the equation of the graph below?

A. $y = 2x + 2$ **B.** $y = x - 2$
C. $y = -x - 2$ **D.** $y = -2x - 2$

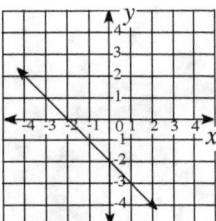

6. Slope, x- and y-intercepts

2–15. Finding the slope of any linear equation.

Any linear equation can be in the form $y = mx + b$, where <u>m is the slope</u> and <u>b is the y-intercept</u>.

2–16. Find the slope and y-intercept.

$$y = 2x + 1$$

SOLUTION

First, look at the equation of $y = 2x + 1$.

The form of the linear equation is:
$y = mx + b$, where m is the slope and b is the y-intercept.

$y = 2x + 1$

So the slope of the line is 2 and the y-intercept is 1. As $m = 2$, that means the linear slope goes upward from left to right.

Quick Exercises 9   Find the slope and y-intercept of each equation.

1) $y = 7x - 2$ **2)** $y = 4x + 1$

2–17. Find the slope and y-intercept.

$$y = -4x + 3$$

SOLUTION

You can easily find the slope and y-intercept.

The form of the linear equation is:
$y = mx + b$, where m is the slope and b is the y-intercept.

$y = -4x + 3$

The slope of the line is −4 and the y-intercept is 3. As $m = -4$, that means the linear slope goes downward from right to left.

Quick Exercises 10   Find the slope and y-intercept.

1) $y = -7x - 2$ **2)** $2y = 4x - 1$

2–18. Find the slope and y-intercept.

$$y = 2x$$

SOLUTION

$\begin{cases} \text{The form of the linear equation is:} \\ y = mx + b, \text{ where } m \text{ is the slope and } b \text{ is the } y\text{-intercept.} \\ \quad\downarrow \\ y = 2x \end{cases}$

So, the slope of the line is 2 and the y-intercept is 0.

Quick Exercises 11   Find the slope and y-intercept.

1) $y = -2x$ **2)** $5y = x - 5$

2–19. How to find the slope given a graph.

To find the slope of a nonvertical line.

$$\text{Slope } (m) = \frac{\text{rise}}{\text{run}}$$

$$m = \frac{y_2 - y_1}{x_2 - x_1}$$

2–20. a) Find the slope and the y-intercept of the line $\overleftrightarrow{AB}$ below. b) Find the linear equation given the information from a).

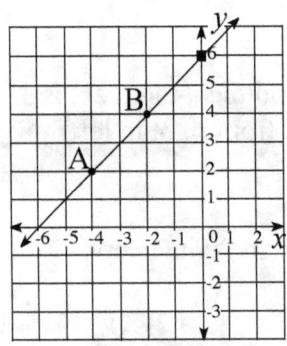

SOLUTION

a) i) Find the slope of the line.
As the line passes through points A(−4, 2) and B(−2, 4), then the slope is:
    Let (x_1, y_1) be A(−4, 2) and (x_2, y_2) be B(−2, 4).

$$m_{AB} = \frac{y_2 - y_1}{x_2 - x_1}$$ Apply the slope formula.

$$m_{AB} = \frac{4 - 2}{(-2) - (-4)} = \frac{2}{2} = 1$$ Substitute in the values of the coordinates.

So, the slope of line m is 1.

ii) To find the y-intercept of a line.
If the line passes above 0 on the x-axis, then the y-intercept of the line is positive.
If the line passes below 0 on the x-axis, then the y-intercept of the line is negative.
The line passes through 6 on the x-axis. So, the y-intercept of the line is 6.

b) The linear equation in the form is $y = mx + b$, where m is the slope and b is the y-intercept. So, the equation of AB is $y = x + 6$.

Quick Exercises 12   Find the slope and y-intercept.

Slope:

y-intercept:

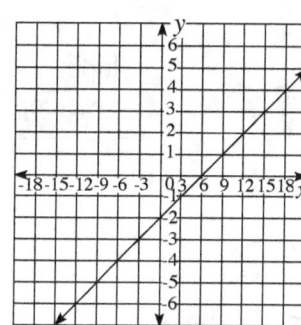

2–21.   Make a linear equation using the given information.

$$m = 2, \ y\text{-intercept} = -1$$

SOLUTION

The linear equation in the form is $y = mx + b$, where m is the slope and b is the y-intercept. So, the equation is $y = 2x - 1$.

Exercises 33   Find the slope and y-intercept.

1) $y = 3x - 1$

Slope:
y-intercept:

2) $y = 3x - \dfrac{1}{2}$

Slope:
y-intercept:

3) $y = x + 4$

Slope:
y-intercept:

4) $y = 4x + 3$

Slope:
y-intercept:

5) $y = -7x - 3$

Slope:
y-intercept:

6) $y = -5x$

Slope:
y-intercept:

7) $y = 2x$

Slope:
y-intercept:

8) $y = -1$

Slope:
y-intercept:

9) $y = \dfrac{1}{2}x - 3$

Slope:
y-intercept:

10) $y = -5x + \dfrac{1}{2}$

Slope:
y-intercept:

Exercises 34   Find the linear equation using the given information.

1) $m = 2$, y-intercept $= 1$

2) $m = -2$, y-intercept $= -1$

3) $m = -1$, y-intercept $= 5$

4) $m = 5$, y- intercept $= -2$

5) $m = 3$, y-intercept $= 0$

6) $m = -2$, y-intercept $= 3$

Exercises 35   Find the slope and *y*-intercept of each graph.

1)

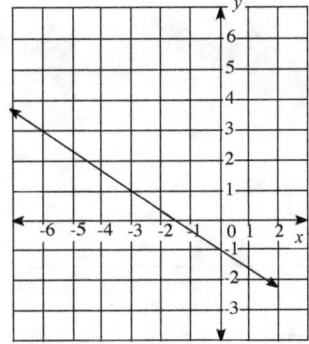

Slope:

y-intercept:

2)

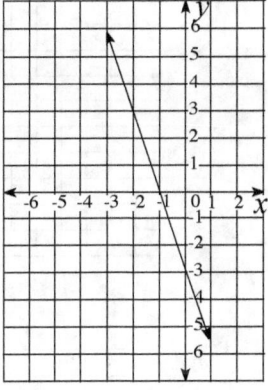

Slope:

y-intercept:

3)

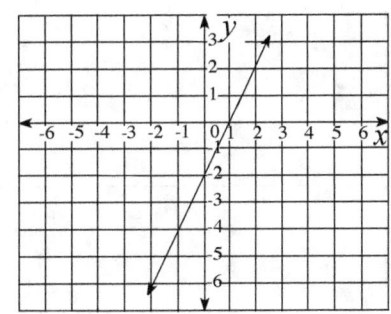

Slope:

y-intercept:

4)

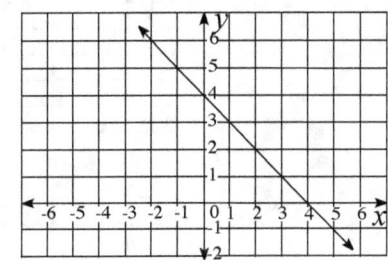

Slope:

y-intercept:

5)

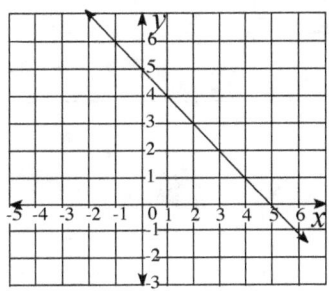

Slope:

y-intercept:

6)

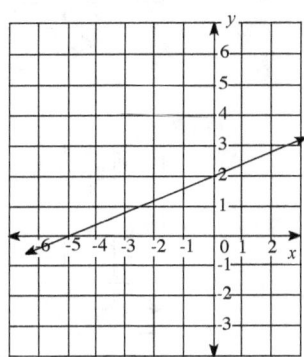

Slope:

y-intercept:

Exercises 36   Use each graph to make an equation.

1)

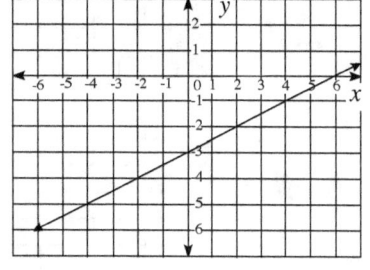

2)

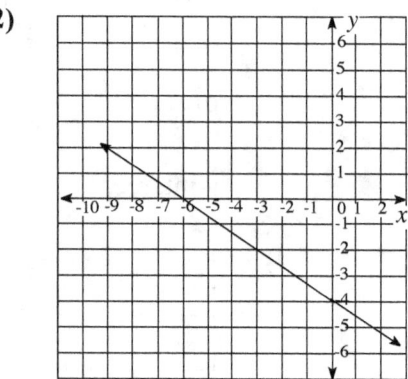

3)

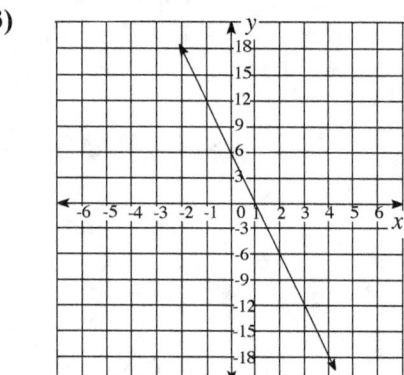

4)

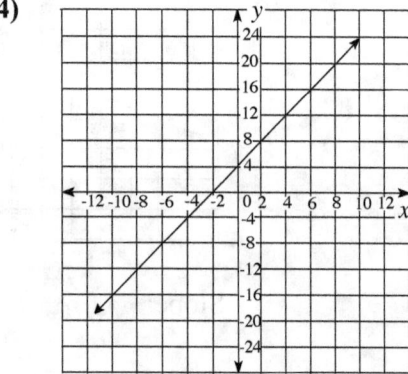

Exercises 37   Find the equation of each graph. Explain how you got your answer.

1)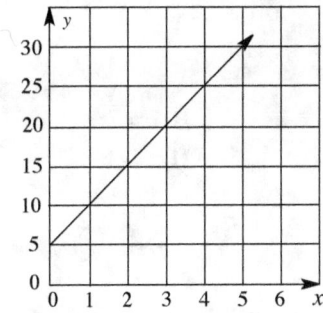

2)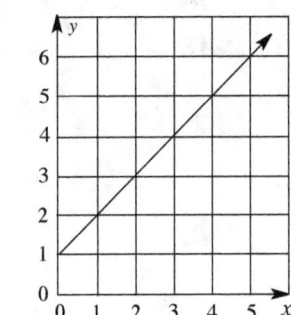

3)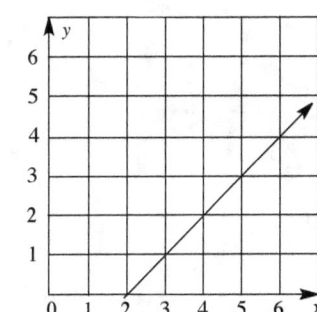

4)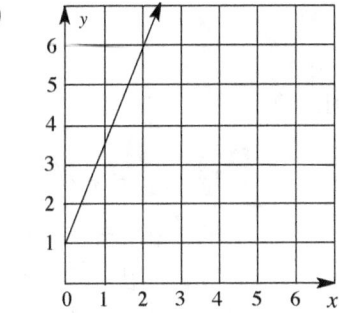

1. What is the slope of $y = -7x + 3$?

   A. 7   B. −7
   C. 3   D. −3

2. What is the y-intercept of $y = \frac{2}{5}x + 3$?

   A. $\frac{2}{5}$ B. $-\frac{2}{5}$
   C. 3   D. −3

3. What is the slope of $3x - y = 1$?

   A. 3   B. −3
   C. 1   D. −1

4. What is the slope of $-2x + 5y = 1$?

   A. 2   B. −2
   C. $\frac{2}{5}$ D. 1

5. What is the y-intercept of $2y - 2x + 4 = 1$?

   A. −3   B. $-\frac{1}{2}$
   C. 1   D. $-1\frac{1}{2}$

6. What is the slope of the line in the graph below?

   A. −2   B. $\frac{1}{4}$

   C. 1   D. $-\frac{1}{4}$

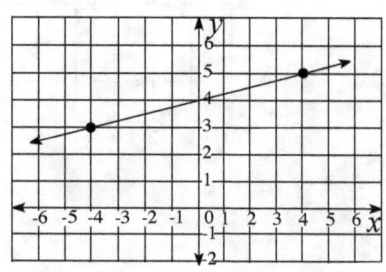

7. What is the slope of the line in the graph below?

A. –2   B. $-\dfrac{1}{2}$

C. –1   D. $-1\dfrac{1}{2}$

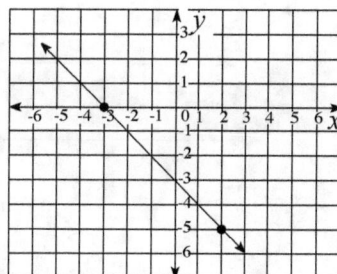

8. What is the slope of the line in the graph below?

A. –3   B. –2

C. 1   D. $-1\dfrac{1}{2}$

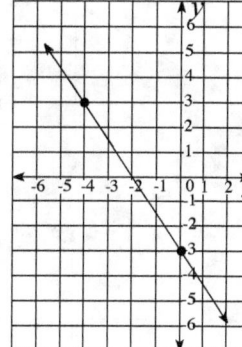

9. What is the *y*-intercept of the line in the graph below?

A. 4   B. –3

C. 1   D. $-2\dfrac{1}{2}$

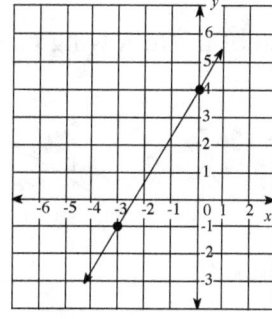

10. Which of the following graphs represents $y = x + 10$?

A. 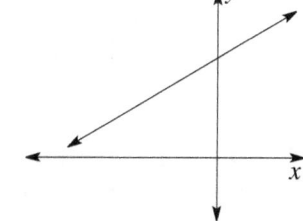   B.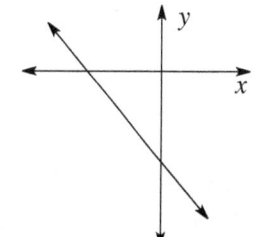

C.

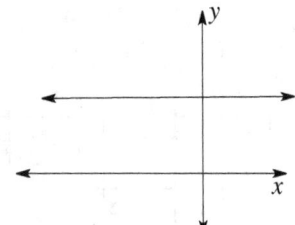

D.

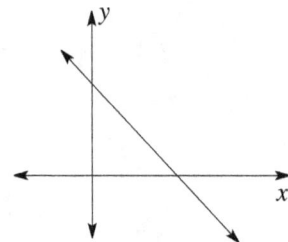

11. Which of the following graphs represents $y = -x - 10$?

A.

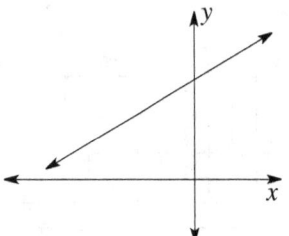

B.

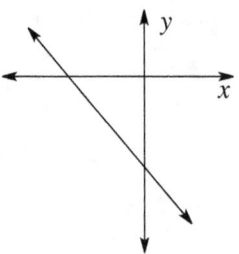

C.

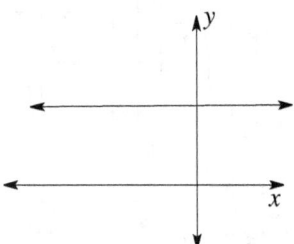

D.

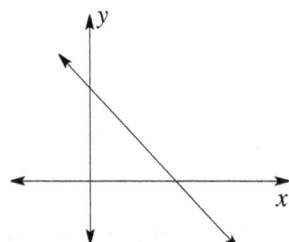

12. Which of the following graphs represents $y = 6$?

A.

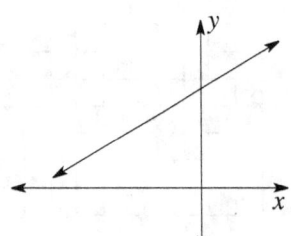

B.

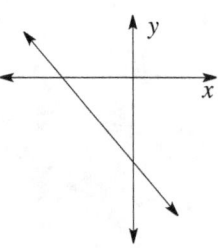

C.

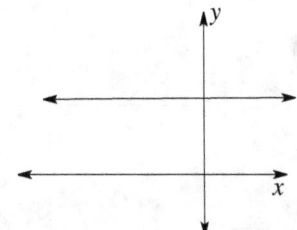

D.

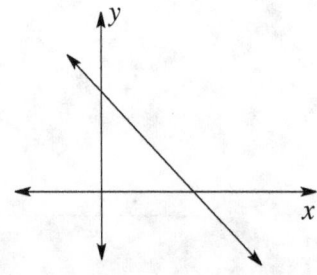

CHAPTER 3
Decimals and Fractions

In this chapter, you will solve problems involving adding, subtracting, multiplying, and dividing decimals and fractions

1. Estimating Sums and Differences with Decimals

3-1. Estimate the sum.

$$(-86.62) + (-12.46)$$

SOLUTION

In addition, if both numbers are negative, then the answer is negative.

(-86.72)
1) Round to the nearest whole number, which is the **6**.
2) Look at the next digit **7**, which is 5 or more.
3) Therefore, **6** will increase by 1.

$+ (-12.46)$
1) Round to the nearest whole number, which is **2**.
2) Look at the next digit **4**, which is less than 5.
3) Therefore, **2** will stay the same.

$$
\begin{array}{r}
(-87) \\
+ (-12) \\
\hline
(-99)
\end{array}
$$

So, the estimated sum of $(-86.62) + (-12.46)$ is -99.

Quick Exercises 1   Estimate the sum.

1) $-4.03 + 2.24$

2) $-5.93 + 7.49$

2. Comparing Decimals

3-2. Comparing the decimals

a. First, compare the signs, the digits, and determine if the place value after the decimal point of both decimals are different from each other.

b. Second, if they are the same number, then compare the next place value with each other and go on until you find a place value where the digits are different.

c. Once you found the different digits, compare them and determine the greater and least of the digits.

3-3. Compare the decimals.

$$1.5 \boxed{} -2.6$$

SOLUTION

When you compare the two numbers, one is a positive number and the other is a negative number. The negative number is smaller than the positive number

So, –2.6 is less than 1.5 or 1.5 is greater than –2.6. Therefore, 1.5 > –2.6.

Quick Exercises 2   Compare the decimals.

1) –0.06 _____ 0.06

2) –5.06 _____ –5.60

Exercises 1   Estimate the sum or difference.

1) $3.75 – $1.20

2)   62.497 – 9.039

_____

3)   632.93 + (–75.39)

4)   4.4 + 11.7

_____

5) $38.46 + (–$84.82)

6)   (–9.49) + (– 5.69)

hint: (+) + (–) = (+) – (+)

hint: (–) + (–) = –

7)   –$19.59 + $53.85

8) $29.49 – $25.79

_____

9)   42.20 – (–5.43)

10)   –12.07 – (–3.48)

hint: (+) – (–) = +

hint: (–) – (–) = (–) + (+)

Exercises 2   Compare the decimals.

1) -0.01 _____ 0

2) 0.92 _____ -1.50

3) 0.42 _____ -0.36

4) -2.03 _____ -3.52

5) -0.74 _____ -0.84

6) -1.06 _____ 0.06

7) -0.29 _____ -0.08

8) -5.06 _____ -5.04

9) -1.06 _____ 0.04

10) $\dfrac{1}{10}$ _____ 0.04

11) 4.075 _____ $4\dfrac{1}{10}$

12) $-\dfrac{8}{100}$ _____ 0.01

Exercises 3   List from greatest to least.

1) $-0.04,\ -0.14,\ 0.04,\ 0.14,\ -1.02$

2) $-0.25,\ -\dfrac{1}{5},\ \dfrac{2}{7},\ -0.30,\ 0$

_____

3) $-\dfrac{4}{7},\ -\dfrac{4}{8},\ -0.625,\ -\dfrac{6}{9}$

_____

4) $-1.03,\ -1\dfrac{2}{10},\ -1.30,\ -1.33,\ -0.99$

_____

_____

5) $-\dfrac{1}{6},\ -\dfrac{1}{8},\ -0.09,\ -\dfrac{1}{9}$

6) $-2\dfrac{1}{4},\ 1\dfrac{1}{3},\ -1.09,\ -1\dfrac{1}{2}$

_____

_____

7) $0.04,\ 0.014,\ 0.14,\ 1.02$

8) $\dfrac{9}{10},\ -\dfrac{8}{100},\ 0,\ -\dfrac{1}{4}$

_____

_____

Exercises 4   Identify the relative position of each value on the number line.

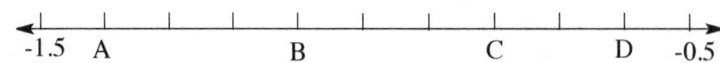

1) A =   **2)** B =
3) C =   **4)** D =

Exercises 5   Identify the relative position of each value on the number line.

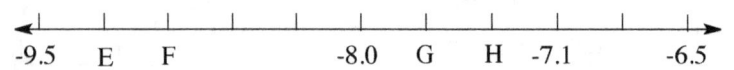

1) E =   **2)** F =
3) G =   **4)** H =

Exercises 6   Identify the relative position of each value on the number line.

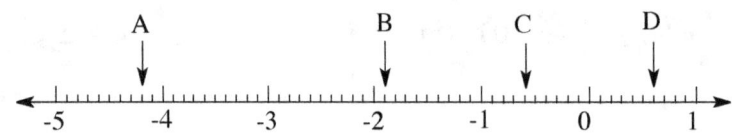

1) A =   **2)** B =
3) C =   **4)** D =

Exercises 7   Identify the relative position of each value on the number line.

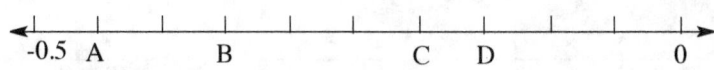

1) A =   **2)** B =
3) C =   **4)** D =

SELF-TEST

1. Which of the following describes the number below in word form?
    820.037

    A. eight hundred twenty and thirty-seven thousand
    B. eight hundred twenty and thirty-seven thousandths
    C. eight hundred twenty, thirty-seven hundredths
    D. eight hundred twenty, thirty-seven thousandths

2. Which of the following describes the expanded form below?
    $$500 + 1 + \frac{2}{10} + \frac{3}{1000}$$

    A. 501.230   B. 510.203
    C. 501.023   D. 501.203

3. What is the estimated difference below?
    $270.05 − $72.91

    A. $196.00   B. $197.00
    C. $197.14   D. $197.20

4. Samantha deposited $408.69 in her bank account one day and $540.95 the next. She then withdraws $650.00 from her bank account to pay a bill. Which of the following is the correct estimate of the money she has remaining?

    A. $299.00   B. $299.64
    C. $300.00   D. $298.00

5. Which of the following is the best estimate of $316.40 + $66.99?

    A. $383.30   B. $382.00
    C. $383.39   D. $383.00

6. Randy put 0.74 pounds of birdseed in the bird feeder. Two hours later, he found that 0.49 pounds of birdseed left in the feeder. Which of the following estimates the number of pounds of birdseed that was eaten during the two hours Randy was gone? Round to the nearest tenth.

    A. 0.2   B. 0.3
    C. 0.4   D. 0.5

7. What is the best estimated sum of $29.09 + $24.61?

    A. $54.00   B. $53.00
    C. $53.60   D. $53.70

8. What is the best estimated difference of $12.55 − $8.49?

    A. $4.06   B. $5.00
    C. $4.00   D. $4.10

9. Will wants to know the perimeter of a rectangle. He measures the length to be 8.52 and the width to be 5.57 inches. Estimate the perimeter to the nearest whole number.

    A. 30   B. 26
    C. 28   D. 54

10. Which of the following is correct?

    A. −0.65 < −0.90   B. −905.99 > 320.55
    C. −0.92 is less than −1.90.   D. 0.2 is greater than 0.02.

11. Which of the following is correct?

    A. 640.638 ≠ six hundred forty and six hundred thirty-eight thousandths
    B. 0.002 is greater than 0.200.
    C. $\dfrac{1}{1000} = 0.01$
    D. 604.050 equals to $604\dfrac{5}{100}$.

Use the number line for Exercises **12-14**.

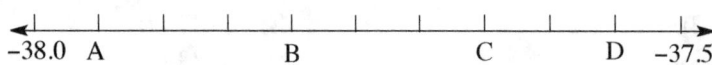

12. What is the relative position of A?

    A. −38.05   B. −38.15
    C. −37.90   D. −37.95

13. What is the relative position of −37.80?
  A. A   B. B
  C. C   D. D

14. Which of the following can round up to −38.00 by the nearest hundredth?

  A. A   B. B and C
  C. C and D   D. B, C, and D

Use the number line for Exercises **15-17**.

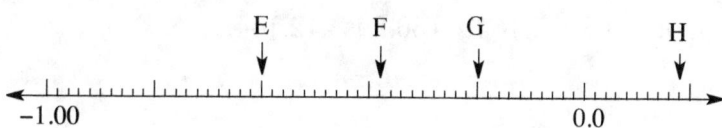

15. What is the relative position of E?

  A. −0.3   B. −0.6
  C. −1.4   D. 0.5

16. Where is the relative position of −0.39?

  A. E   B. F
  C. G   D. H

17. Which of the following can be rounded up to −1.0 by the nearest tenth?

  A. E   B. E and F
  C. E, F, and G   D. H

3. Adding and Subtracting Decimals

3-4. Add the decimals.

$$49.08 + 2.164 + 160.9$$

SOLUTION

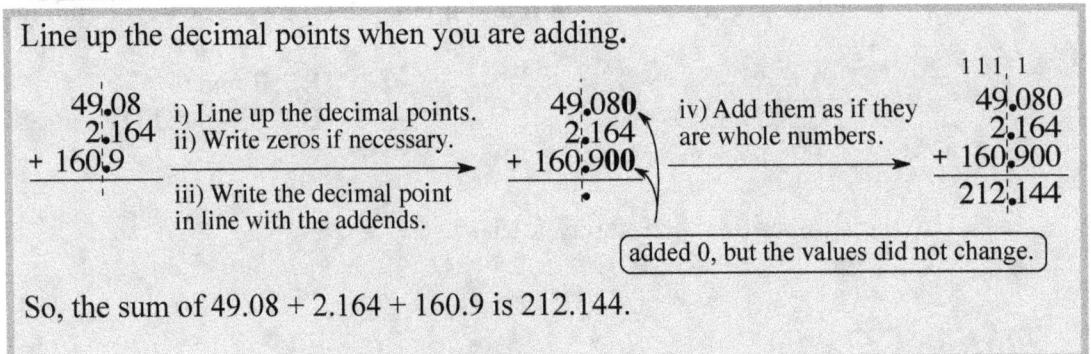

Line up the decimal points when you are adding.

i) Line up the decimal points.
ii) Write zeros if necessary.
iii) Write the decimal point in line with the addends.
iv) Add them as if they are whole numbers.

added 0, but the values did not change.

So, the sum of $49.08 + 2.164 + 160.9$ is 212.144.

Quick Exercises 3   Add the decimals.

1) $0.20 + (-0.9)$ **2)** $-5.6 + (-2.09)$

3-5. Subtract the decimals.

$$(-19.02) - (-7.6)$$

SOLUTION

a. Line up the decimal points when you are subtracting.

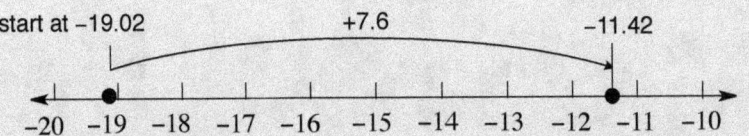

start at –19.02   +7.6   –11.42

Start at -19.02 on the number line and as $- (-7.6) = +7.6$, moves 7.6 places to the right.

b. Change the operation sign if the second number is –(–), then always change it into a positive number.

i) Line up the decimal points.
ii) Change –(–) into +.
iii) Write zero if necessary.
iv) Write the decimal point in line with the addends.
iv) Solve the problem.

added 0, but the value did not change.

So, the difference of $(-19.02) - (-7.6)$ is -11.42.

3–6. Subtract the decimals.

$$25.432 - 7.66$$

SOLUTION

Line up the decimal points when you are subtracting.

$\begin{array}{r} 25{.}432 \\ -\ 7{.}66 \\ \hline \end{array}$

i) Line up the decimal points.
ii) Write zeros if necessary.

→

$\begin{array}{r} 25{.}432 \\ -\ 7{.}660 \\ \hline \end{array}$

iii) Write the decimal point in line with the addends.

iv) Subtract them as if they are whole numbers.

→

$\begin{array}{r} \overset{14\ 13}{\cancel{15}}.\overset{1}{\cancel{4}}\overset{3}{\cancel{3}}13 \\ 25{.}432 \\ -\ 7{.}660 \\ \hline 17{.}772 \end{array}$

added 0, but the value did not change.

So, the difference of $25.432 - 7.66$ is 17.772.

Quick Exercises 4   Solve each expression.

1) $2.03 - (-3.9)$

2) $-0.6 - (-0.09)$

Exercises 8   Add the decimals.

1) $\$18.25 + (-\$13.77)$

2) $(-\$2.30) + \$1.46$

3) $(-2.72) + (-5.24)$

4) $(-1.2) + 2.4$

5) $5.03 + (-7.55)$

6) $92.09 + (-54.74)$

7) $35.43 + (-98.08)$

8) $-41.73 + 26.05$

Exercises 9   Add the decimals.

1) $0.253 + (-1)$

2) $(-0.253) + 0.056$

3) $(-0.646) + (-0.552)$

4) $(-0.606) + 0.597$

5) $0.053 + (-1.05)$

6) $0.056 + (-0.09)$

7) $\$45.65 + (-\$98.08) + \$0.93$

8) $-24.073 + 18.059 + 9.24$

9) $25.089 + 19.936 + (-18.2)$

10) $-\$38.57 + \$75.99 + \$20.55$

11)

$$\begin{array}{r} 20.02 \\ + \ 2.09 \\ \hline \end{array}$$

12)

$$\begin{array}{r} 19.056 \\ + \ 13.68 \\ \hline \end{array}$$

13)

$$\begin{array}{r} 18.44 \\ + \ 17.582 \\ \hline \end{array}$$

Exercises 10   Find each value.

1)   If $x = 0.25$, find the value of $-0.84 + x$.

2)   If $x = -1.83$, find the value of $x - 2.36$.

3)   If $-x = 1.21$, find the value of $1.68 - x$.

4)   If $-x = -5.13$, find the value of $x - 3.65$.

5)   If $x - 2.04 = 1.08$, find the value of $x - 1.08$.

6)   If $2(-0.29 + x) = 0.76$, find the value of $x - 0.38$.

Exercises 11   Estimate the difference of the decimals.

1) $63.3 - (-28.09)$

2) $-0.56 - (-0.05)$

3) $-2.27 - (-4.88)$

4) $-1.366 - 4.07$

5) $(-0.502) - (-0.26)$

6) $2.042 - (-1.82)$

7) $101.68 - 27.077$

8) $41.72 - 13.47$

9)
$$\begin{array}{r} 43.05 \\ -\ 14.051 \\ \hline \end{array}$$

10)
$$\begin{array}{r} 17.933 \\ -\ 15.07 \\ \hline \end{array}$$

11)
$$\begin{array}{r} 61.93 \\ -\ 15.025 \\ \hline \end{array}$$

Exercises 12   Find the value using the given information.

1)   If $y - 3 = -0.2$, find the value of $0.7(y - 3)$.

2)   If $\dfrac{7}{10} - 2y = -2.73$, find the value of $0.7 - 2y$.

3)   If $2(3.2 + x) = -4$, find the value of $(-3.2 - x)$.

4)   If $-(x - 1.8) = -2$, find the value of $2 - x$.

5)   If $\dfrac{x - 1}{3} = 0.2$, find the value of $2(x - 1)$.

6)   If $\dfrac{1 - x}{5} = 0.4$, find the value of $1 - (x - 2)$.

Exercises 13   Subtract the decimals.

1) $(-0.52) - (-0.966)$

2) $0.602 - 0.499$

3) $10.502 - (-4.53)$

4) $(-6) - (-4.762)$

5) $18.06 - 20.29$

6) $7.50 - 8\%$

7) $(-52.42) - (-32.26)$

8) $38\% - 10.42$

Exercises 14   Find each value of Δ.

1) $2.03 - \Delta = 1.89$

2) $0.79 + \Delta = 2.01$

3) $\Delta + 11.35 = 17.29$

4) $\Delta - 1.81 = 3.14$

5) $16.09 + \Delta = 19.81$

6) $\Delta + 0.42 = 1.51$

7) $25.01 - \Delta = 13.63$

8) $4.33 - \Delta = 1.90$

9) $\Delta + 2.9 = 6.92$

10) $\Delta - (-9.05) = 6.08$

11) $\$6.32 - (-\Delta) = \$1.08$

12) $\Delta + (-\$2.55) = \$10.00$

* Solving Problems

Exercises 15   Solve each problem using the given information.

1) At a fundraiser, 15 people each donated the same amount of money. If the money they donated adds up to $187.95, what is the equation that shows how much money each person donated?

2) The lengths of the sides of a triangle are 9.26 ft, 9.26 ft, and 6.72 ft. What is the perimeter of the triangle?

3) Carol has $61.29. She wants to buy her friend a present that costs $45.85. Additionally, she bought a muffler for herself that cost $13.75. Both items are 20% off. How much money will she have left?

SELF-TEST

1. What is the value of the equation below?
$$(-8.35) - (-29.49)$$

   A. −21.14   B. −37.84
   C. 21.14   D. 37.84

2. Mrs. Gerald is baking cookies. She usually puts in 3.5 cups of sugar for a batch of cookies. If Mrs. Gerald decides to quadruple the recipe, how many cups of sugar does she need?

   A. 3.5   B. 10.5
   C. 0.875   D. 14

3. Which of the following is the value of $25.49 + (-54.55)$?

   A. −29.06   B. 80.04
   C. 29.06   D. −80.04

4. Bryan made 1.5 gallons of lemonade. He then spilled half of the lemonade he made.

Afterwards, he fills 5 glasses with equal amounts of lemonade. How much lemonade would be in each glass?

A. 0.75   B. 0.45
C. 0.30   D. 0.15

5. Daniel is twice as tall as Pierce. Pierce is 80.25 cm tall. How tall is Daniel rounded to the nearest tenth?

A. 40.125 cm   B. 80.25 cm
C. 120.375 cm   D. 160.500 cm

6. What is the sum of $-\$307.05 + \$489.26$?

A. $-\$182.21$ B. $\$182.21$
C. $\$489.26$ D. $\$796.31$

7. It rained 0.52 inches on Monday. On Tuesday, it rained 0.19 inches greater than on Monday. How much did it rain in total?

A. 0.19   B. 0.71
C. 1.04   D. 1.23

8. Given that Kim has $854.35 in her bank account and withdraws $65.00 from her bank account, how much does he have left in her account?

A. $854.35   B. $65.00
C. $789.35   D. $919.35

9. What is the value of Δ for the equation below?
$$\$73.05 - \Delta = -\$25.80$$

A. $47.25   B. $-\$98.85$
C. $98.85   D. $-\$25.80$

10. Corby and his friends went to a restaurant. They paid $124.30 for their meal. How many people went to the restaurant if each person paid $24.86?

A. 3   B. 4
C. 5   D. 6

4. Multiplying and Dividing Decimals

3–7. For multiplying decimals.

> a) Multiply normally, ignoring the decimal points.
> b) In an expression, if both numbers are negative, then the answer is always positive. If both numbers are positive, then the answer is always positive. And if one of the numbers is negative, then the answer will be always negative.
> c) Place the decimal point in the answer- it will have as many decimal places as the two original numbers combined.

3–8. Multiply the decimals.

$$(-5.6) \times 7$$

SOLUTION

In the expression, one of the numbers is negative.

$$
\begin{array}{r} (-5.6) \\ \times \ 7 \\ \hline \end{array} \longleftarrow 1 \text{ decimal place}
\qquad \text{i), ii)} \Longrightarrow
\overset{4}{\begin{array}{r} (-5.6) \\ \times \ \ 7 \\ \hline -392 \end{array}}
\qquad \text{ii)} \Longrightarrow
\overset{4}{\begin{array}{r} (-5.6) \\ \times \ \ 7 \\ \hline -39.2 \end{array}}
\left. \right\} 1 \text{ decimal place}
$$

1 decimal place

i) Multiply like whole numbers.
ii) In multiplication, if one of the numbers is negative, then the answer is negative.
iii) When you find the product, move **one** decimal place to the left.

So, the product of $(-5.6) \times 7$ is -39.2.

* When multiplying with negatives.

In multiplication:	
$(-) \times (-) = +$	$(-) \times (+) = -$
$(+) \times (+) = +$	$(+) \times (-) = -$

Quick Exercises 5   Solve each expression.

1) $(-1.5) \times 2.5$

2) $(-0.74) \times 1.2$

3) 0.38×4

4) $0.03 \times (-7)$

3-9. Multiply the decimals.

$$21.85 \times 6$$

SOLUTION

In the expression, one of the numbers is a decimal.

$$21.85 \leftarrow \textbf{2} \text{ decimal places}$$
$$\times \quad 6$$

i)
$$\overset{1\,5\,3}{21.85}$$
$$\times \quad 6$$
$$\overline{13110}$$

ii)
$$21.85$$
$$\times \quad 6$$
$$\overline{131.10} \Big\} \textbf{2} \text{ decimal places}$$

$2 \; 1$ **2** decimal places

i) Multiply like whole numbers.

ii) When you find the product, move **two** decimal places to the left.

So, the product of 21.85×6 is 131.10.

Quick Exercises 6   Solve each expression.

1) $(-16.3) \times 2$ **2)** $(-3.5) \times (-1.4)$

3) $(-2) \times 0.37$ **4)** $(-6.2) \times (-6)$

3-10. Multiply the decimals.

$$5.106 \times 2.2$$

SOLUTION

In the expression, one of the numbers is a decimal.

$$5.106 \leftarrow \textbf{3} \text{ decimal places}$$
$$\times \quad 2.2 \leftarrow \textbf{1} \text{ decimal place}$$

i)
$$\overset{1}{\overset{1}{5.106}}$$
$$\times \quad 2.2$$
$$\overline{10212}$$
$$+ 10212$$
$$\overline{112332}$$

ii)
$$5.106$$
$$\times \quad 2.2 \Big\} \textbf{4} \text{ decimal places}$$
$$\overline{10212}$$
$$+ 10212$$
$$\overline{11.2332}$$
$4\;3\;2\;1$ **4** decimal places

i) Multiply like whole numbers.

ii) When you find the product, move **four** decimal places to the left.

So, the product of 5.106×2.2 is 11.2332.

Quick Exercises 7   Solve each expression.

1) 0.693×2

2) $(-4.006) \times (-2)$

3) $0.023 \times (-5)$

4) $(-0.06) \times 1.2$

3-11.   For dividing decimals.

> a) First, move the decimal places for both numbers until the divisor is a whole number.
> b) Line up the decimal point with the dividend.
> c) In an expression, if both numbers are negative, then the answer is always positive. If both numbers are positive, then the answer is always positive. And if one of the numbers is negative, then the answer will be always negative.

3-12.   Dividing the decimals.

$$(-2.286) \div 0.3$$

SOLUTION

a) First, move **one** decimal place for both numbers until the divisor is a whole number.

i) Move one decimal place on both.

$0.3\overline{)(-)2.286}$ $\longrightarrow$ $3\overline{)(-)22.86}$

* So this means that the expression of $-2.286 \div 0.3$ is equivalent to $-22.86 \div 3$.

b) Second, line up the decimal point with the dividend.
c) In expression, if one of the numbers is negative, then the answer is negative.

$3\overline{)(-)22.86}$

ii) Line up the decimal points.
iii) If one of the numbers is negative, then the answer is negative.

$\longrightarrow$ $3\overline{)(-)22.86}$ (-) •

d) Now divide.

$3\overline{)(-)22.86}$ (-) •

iv) Solve like whole numbers.

$\longrightarrow$

```
      (-) 7.62        divisor
  3)(-)22.86       quotient
    - 21           3 x 7 = 21
    ─────
      18                 dividend
    - 18           3 x 6 = 18
    ─────
      06
     - 6           3 x 2 = 6
    ─────
       0
```

So, the quotient of $(-2.286) \div 0.3$ is (-7.62).

3–13. Divide the decimals.

$$10.268 \div 3.02$$

SOLUTION

a) First, move the decimal point to the right for both numbers until the divisor is a whole number.

$$3.02\overline{)10.268} \quad\longrightarrow\quad 302\overline{)1026.8}$$

Move the decimal point two places to the right for both numbers

b) Second, line the decimal point in the quotient with the decimal point of the dividend.

$$302\overline{)1026.8} \quad\longrightarrow\quad 302\overline{)1026.8}$$

c) Now divide

$$
\begin{array}{r}
3. \\
302\overline{)1026.8} \\
-906 \\
\hline
120
\end{array}
\quad 302 \times 3 = 906
\quad\longrightarrow\quad
\begin{array}{r}
3.4 \\
302\overline{)1026.8} \\
-906 \\
\hline
1208 \\
-1208 \\
\hline
0
\end{array}
\quad \begin{array}{l} 302 \times 3 = 906 \\ \\ 302 \times 4 = 1208 \end{array}
$$

So, the quotient of $10.268 \div 3.02$ is 3.4.

* When dividing with negatives.

In division:

$(-) \div (-) = +$ $(-) \div (+) = -$

$(+) \div (+) = +$ $(+) \div (-) = -$

Quick Exercises 8 Solve each expression.

1) $(-6.459) \div 3$ 2) $(-21.728) \div (-7)$

3) $\dfrac{3}{4} \div (-3)$ 4) $2.5 \div (-10)$

Exercises 16   Find the product of the decimals.

1) $(-2.50) \times 8$

2) $(-0.47) \times 2$

3) 0.3×2.6

4) $7.2 \times (-0.5)$

5) $(-5.05) \times \dfrac{7}{10}$

6) $(-2.08) \times (-6)$

7) $8\% \times 15$

8) $25 \times 20\%$

Exercises 17   Find the product of the decimals.

1) $\$12.89 \times \$8.28$

2) $6.3 \times (-4)$

3) $2.54 \times (-9)$

4) $(-3.8) \times (-5)$

5) $(-8.2) \times (-3)$

6) $(-6) \times (-2.4)$

7) $(-0.25) \times (-0.9)$

8) $(-5.04) \times (-6.2)$

9) $(-4.1) \times 13\%$

10) $3\% \times (-3.2)$

Exercises 18   Find the value using the given information.

1)   If $y = -3.8$, find the value of $y \times 1.5$.

2)   If $-y = 0.4$, find the value of $2.05 \times y$.

_____

3)   If $-x = -0.5$, find the value of $2.5 \times x$.

4)   If $x = -1.5$, find the value of $x \times (-2.6)$.

_____

5)   If $y \times 0.25 = 1.85$, find the value of $y \times (-0.42)$.

6)   If $0.34 \div y = -3.6$, find the value of $3.6 \times y$.

_____

Exercises 19   Multiply the decimals.

1) 4.8×2.3

2) 0.42×0.21

_____

3) $(-2.50) \times 1.2$

4) $(-0.64) \times 2.9$

_____

5) $91.62 \times (-3.56)$

6) $5.22 \times (-3.5)$

_____

7) $0.007 \times 1.5 + (-1)$

8) $\$8.63 \times \$9.08 - \$52.35$

_____

9)
$$52.08$$
$$\times\ 20.5$$

10)
$$16.04$$
$$\times\ 17.99$$

11)
$$28.84$$
$$\times\ 16.93$$

Exercises 20   Find the value of x.

1) $2.03 - 2x = 1.89$

2) $0.78 + 3x = 2.01$

3) $3x + 11.35 = 17.29$

4) $5x - 1.81 = 3.14$

5) $2(3.2 + x) = -4.48$

6) $4(x - 0.42) = 1.64$

7) $24.97 - 9x = 13.63$

8) $4.05 - 4x = 1.65$

9) $0.5(x + 2.9) = 6.92$

10) $0.8(x - 9.05) = 6.08$

Exercises 21   Find the value using the given information.

1)   If $y + (-2.21) = 0.89$, find the value of $y \times 10$.

2)   If $-y + 8.32 = 13.62$, find the value of $3(-4 \times y)$.

3)   If $\frac{5}{6}x = -7.5$, find the value of $(0.5 \times x)$.

4)   If $4x = -8.52$, find the value of $x \times (-20)$.

5)   If $\frac{1}{2}(12 - 4.8) = 2y$, find the value of $y \times (-4.2)$.

6)   If $(0.25 + 1.05) \div y = -2.1$, find the value of $2.1 \times y$.

Exercises 22   Find the quotient.

1) $9.3 \div 0.5$

2) $13.65 \div 2.1$

3) $29.12 \div 5.2$

4) $0.119 \div 0.7$

5) $(-5.06) \div 4.6$

6) $(-42.75) \div 9.5$

7) $1.840 \div (-0.2)$

8) $0.78 \div (-0.6)$

9) $21.75 \div 2.9$

10) $(-0.528) \div 2.2$

11) $(-3.24) \div (-7.5)$

12) $(-36.04) \div (-6.8)$

Exercises 23   Find the quotient.

1) $(-0.2952) \div 1.23$

2) $(-2.928) \div 0.48$

3) $7.35 \div (-0.03)$

4) $1.541 \div (-0.23)$

5) $(-2.88) \div (-0.64)$

6) $(-0.91) \div (-0.14)$

7) $36.25\% \div (-1.45)$

8) $2.1\% \div (-0.04)$

Exercises 24   Find the value using the given information.

1) If $y = -3.75$, find the value of $y \div 1.5$.

2) If $-y = 0.8$, find the value of $2.08 \div y$.

3) If $-x = -0.5$, find the value of $2.5 \div x$.

4) If $5.4 + y = 1$, find the value of $(y \div 1.1)^2$.

5) If $(y \div 3.87) = 0.5$, find the value of $2(y \div 3.87)$.

6) If $1.6 \div y = -0.3$, find the value of $2(1.6 \div y)^2$.

Exercises 25   Find the value of x.

1) $7.224 \div x = 6.02$

2) $0.306 \div x = 0.34$

3) $x \div 2.4 = 0.07$

4) $x \div (-5.25) = 3.60$

5) $16.74 \div 2x = 2.7$

6) $6x \div 0.12 = 1.5$

7) $6.18 \div 0.5x = -4.12$

8) $0.22 \div 2.2x = 0.25$

9) $3(x \div 1.4) = 4.8$

10) $-0.25(9.84 \div x) = 1$

11) $2(7.5 \div x) = -1.2$

12) $\frac{1}{2}(x \div 2.3) = 1.87$

* Solving Problems

Exercises 26   Solve each problem using the given information.

1) Daniel is measuring his cell phone and finds that the diagonal of the phone is
    4.5 inches and the length is 3.55 inches. What is the width of the cell phone?
    Show your work and round to the nearest hundredth if necessary.
    What is the perimeter of the cell phone? Show your work.

2) If a cardboard box has a volume of 235.683 cm^3, a width of 4.3 cm, and a
    height of 8.7 cm, what is the length? Show your work.

3) A table is set out for a buffet. On the table are a few plates of crab. Each plate
    has exactly 0.45 pounds of crab. There are 18.675 pounds of crab in total.
    What is the equation to find how many plates of crab are set out? How many
    plates of crab are there? Show your work.

Exercises 27   Solve each problem using the given information.

* CK bought 11.24 lb of pre-cooked ham for Thanksgiving dinner. Before serving,
  she baked it at 275 °F until heated through. The ham needs to be cook 14 minutes
  for every pound.
1) If she paid $27.99 for the ham, how much money did she pay per pound?
    Round to the nearest hundredth.

2) How many hours did she cook the ham?

3) If she served only three quarters of the ham for dinner, how many pounds are
    left?

SELF-TEST

1. Find the sum of the expression below.
$$(-4.005) + 4.405$$

   A. 4.004   B. 0.4
   C. 8.41   D. 0.405

2. Which of the following is the product of $(-2.736) \times 23.2$?

   A. −8.4795   B. −63.4752
   C. 8.4795   D. 63.4752

3. What is the product of the expression below?
$$(-3.07) \times (-2.53)$$

   A. 5.6   B. 7.7671
   C. 0.54   D. −7.7671

4. Find the quotient of the expression below.
$$1.3464 \div 3.6$$

   A. 0.374   B. 2.2536
   C. −0.374   D. 4.9464

5. The cost of 7 books is $110.25. What is the price of a single book?

   A. −$15.75   B. $771.75
   C. $15.75   D. $103.25

6. Find the difference of the expression below.
$$(-4.38) - (-12.93)$$

   A. −8.55   B. 2.952
   C. 8.55   D. 17.31

7. Bess wants to buy a book that cost $23.49. If she paid $25.60 for the book, what is the tax percentage paid?

A. 8.5%	**B.** 3.5%
C. 9%	**D.** 9.5%

8. C.J. spent $69.75 to buy movie tickets. How many tickets did he buy if a ticket cost $13.95?

A. 3	**B.** 4
C. 5	**D.** 6

9. What is the value of Δ for the equation below?

$$\$6.34 \times \Delta = \$53.89$$

A. $8.50	**B.** $60.23
C. $341.70	**D.** $347.55

10. What is the value of Δ for the equation below?

$$\$73.87 \div \Delta = 8.9$$

A. $82.77	**B.** $8.30
C. $64.97	**D.** $657.44

11. What is the value of Δ for the equation below?

$$\Delta \div 5.04 = 12.95$$

A. 2.57	**B.** 7.91
C. 17.99	**D.** 65.268

12. What is the value of Δ for the equation below?

$$\Delta \times 8.5 = 45.90$$

A. 5.40	**B.** 54.40
C. 37.40	**D.** 390.15

5. Comparing Fractions

3–14. Comparing Fractions

> If two different fractions have the <u>same denominators</u>, then you can compare the numerators in order to compare the fractions themselves.

3–15. Compare the fractions.

$$\frac{5}{9} \ \square \ \frac{8}{9}$$

SOLUTION

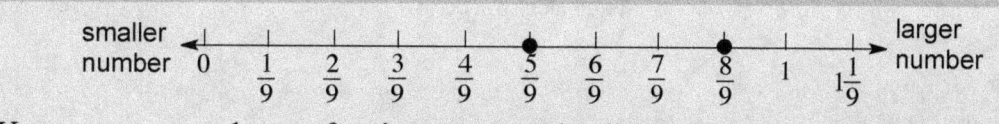

You can compare the two fractions on a number line to determine which fraction is greater.

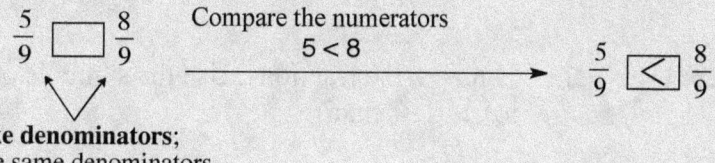

3–16. If two different fractions have <u>different denominators</u>, then follow the directions to compare the fractions.

> a. First, find the LCM of the denominators of the fractions.
> b. Second, rewrite the fractions as equivalent fractions.
> c. Then compare the numerators of the fractions.

3–17. Compare the fractions.

$$\frac{3}{4} \ \square \ \frac{4}{9}$$

SOLUTION

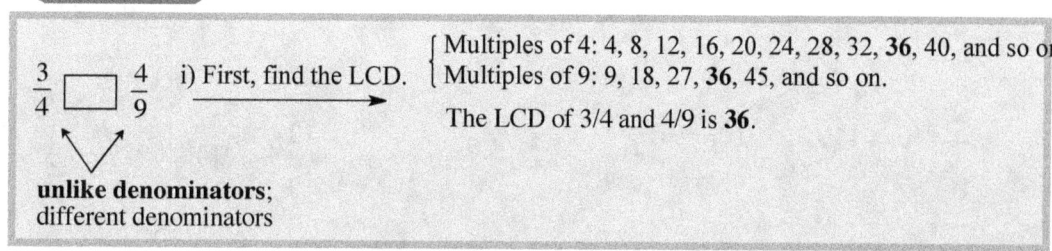

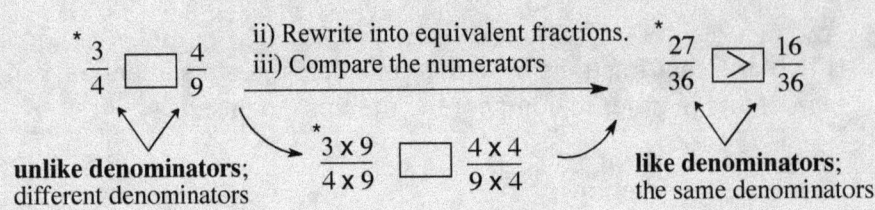

You have to find like denominators and then compare the two different fractions.

*$\dfrac{3}{4}$ ☐ $\dfrac{4}{9}$ ii) Rewrite into equivalent fractions.   *$\dfrac{27}{36}$ ▷ $\dfrac{16}{36}$
iii) Compare the numerators

*$\dfrac{3 \times 9}{4 \times 9}$ ☐ $\dfrac{4 \times 4}{9 \times 4}$

unlike denominators;
different denominators

like denominators;
the same denominators

* Equivalent fractions: The fractions are the same.

$$\dfrac{3}{4} \;=\; \dfrac{3 \times 9}{4 \times 9} \;=\; \dfrac{27}{36}$$

equivalent fractions

Exercises 28   Compare the fractions. Use the symbols of < (greater than), = (equal to), > (less than).

1) $\dfrac{2}{5}$ ☐ $\dfrac{3}{8}$ 2) $-\dfrac{3}{4}$ ☐ $\dfrac{5}{7}$

3) $\dfrac{3}{6}$ ☐ $\dfrac{4}{5}$ 4) $\dfrac{4}{3}$ ☐ $-\dfrac{7}{10}$

5) $\dfrac{8}{9}$ ☐ 1.25 6) 0.95 ☐ $\dfrac{9}{10}$

7) 2.15 ☐ $3\dfrac{1}{4}$ 8) $-\dfrac{1}{3}$ ☐ $-\dfrac{1}{4}$

9) $-\dfrac{5}{6}$ ☐ $-\dfrac{6}{7}$ 10) $-\dfrac{1}{4}$ ☐ 0.05

11) $-1\dfrac{1}{2}$ ☐ $-1\dfrac{1}{3}$ 12) 0.12 ☐ $-1\dfrac{3}{4}$

Exercises 29   List from greatest to least.

1) $\frac{1}{8}$, $\frac{2}{3}$, $\frac{3}{4}$, $\frac{1}{2}$

2) -0.6 $\frac{2}{3}$, $\frac{4}{7}$, $\frac{5}{8}$

3) $\frac{1}{3}$, $\frac{3}{5}$, $\frac{3}{4}$, $\frac{1}{2}$

4) 0.375, $\frac{2}{5}$, $\frac{5}{8}$, 0.309

5) 0.25, $\frac{2}{5}$, $\frac{1}{3}$, -0.30

6) -1.25, $1\frac{5}{16}$, 1.25, $-1\frac{1}{9}$

7) $-\frac{1}{3}$, $-\frac{1}{4}$, $-\frac{1}{6}$, $-\frac{1}{5}$

8) $-1\frac{1}{3}$, $-4\frac{1}{6}$, $-3\frac{1}{5}$, $-2\frac{1}{4}$

Exercises 30   Identify the relative position of each value on the number line.

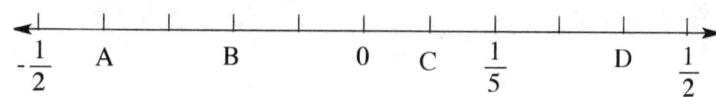

1) A =
2) B =
3) C =
4) D =

Exercises 31   Identify the relative position of each value on the number line.

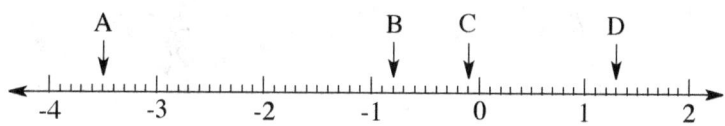

1) A =
2) B =
3) C =
4) D =

Exercises 32   Identify the relative position of each value on the number line.

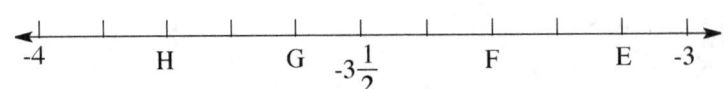

1) E =   2) F =
3) G =   4) H =

Exercises 33   Identify the relative position of each value on the number line.

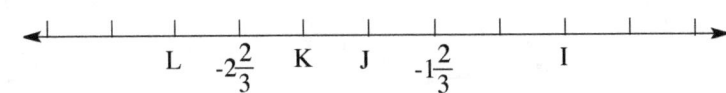

1) I =   2) J =
3) K =   4) L =

SELF-TEST

1. Which of the following is the prime factorization of 48?

    A. 6×8 B. 2×14
    C. $(4)^2 \times 3$ D. $(2)^4 \times 3$

2. What is the LCD of $\frac{2}{5}$ and $\frac{2}{3}$?

    A. 2   B. 4
    C. 5   D. 15

3. Which of the following is the prime factorization of 54?

    A. $2 \times 3 \times 7$ B. 2×3^3
    C. 6×9 D. $2(27)$

4. Find the LCM of 2 and 5.

    A. 2   B. 5
    C. 7   D. 10

5. Which of the following is the GCF of 12 and 8?

    A. 2   **B.** 4

    C. 8   **D.** 12

6. What is the LCD of $1\frac{2}{6}$ and $\frac{1}{9}$?

    A. 6   **B.** 9

    C. 18   **D.** 54

7. Which of the following is correct?

    A. $\frac{5}{6} < \frac{6}{9}$ **B.** $\frac{5}{6} > \frac{6}{9}$

    C. $\frac{5}{6}$ is less than $\frac{6}{9}$ **D.** $\frac{6}{9}$ is greater than 1

8. Which of the following is correct?

    A. $\frac{3}{7} \neq$ three out of seven   **B.** $\frac{1}{10}$ is greater than $\frac{9}{100}$.

    C. $\frac{9}{100} = 0.009$ **D.** 2.49 equals to $2\frac{49}{1000}$.

Use the number line for Exercises **9-11**.

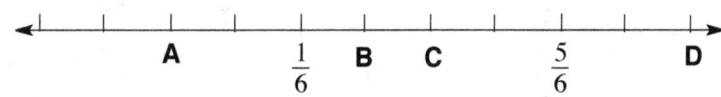

9. What is the relative position of **C**?

    A. $\frac{1}{3}$ **B.** $1\frac{1}{6}$

    C. 0.5   **D.** $\frac{2}{3}$

10. Which of the following is the relative position of $\frac{1}{3}$?

A. A   B. B
C. C   D. D

11. Which of the following can be 1, rounded to the nearest tenth?

A. A   B. B
C. C   D. D

12. Which of the following is correct?

A. $\frac{5}{12} < \frac{4}{10}$ B. $\frac{2}{4} < \frac{3}{6}$

C. $\frac{1}{10} < \frac{1}{4}$ D. $\frac{1}{2} < \frac{1}{9}$

13. Which of the following is correct?

A. $-\frac{1}{3} > \frac{2}{5}$ B. $-\frac{4}{5} > \frac{2}{5}$

C. $\frac{1}{3} < \frac{1}{4}$ D. $\frac{1}{3} < \frac{2}{5}$

14. Which of the following correctly lists the numbers below from least to greatest?
$$\frac{1}{4}, \frac{2}{3}, -\frac{6}{7}, \frac{4}{9}, \frac{3}{5}$$

A. $-\frac{6}{7} > \frac{2}{3} > \frac{3}{5} > \frac{4}{9} > \frac{1}{4}$ B. $-\frac{6}{7} < \frac{1}{4} < \frac{3}{5} < \frac{2}{3} < \frac{4}{9}$

C. $-\frac{6}{7} < \frac{1}{4} < \frac{3}{5} < \frac{4}{9} < \frac{2}{3}$ D. $-\frac{6}{7} < \frac{1}{4} < \frac{4}{9} < \frac{3}{5} < \frac{2}{3}$

15. Which of the following correctly lists the numbers below from least to greatest?
$$0.625, \frac{5}{12}, -\frac{9}{10}, -0.926, \frac{4}{6}$$

A. $-0.926 < -\frac{9}{10} < 0.625 < \frac{5}{12} < \frac{4}{6}$ B. $-0.926 < -\frac{9}{10} < \frac{5}{12} < 0.625 < \frac{4}{6}$

C. $-0.926 < -\frac{9}{10} < \frac{5}{12} < \frac{4}{6} < 0.625$ D. $-\frac{9}{10} < -0.926 < \frac{5}{12} < 0.625 < \frac{4}{6}$

6. Estimating Fractions

3–18. Estimate.

$$2\frac{3}{4} + 3\frac{4}{9}$$

SOLUTION

There are two mixed numbers with whole numbers and fraction numbers. The fraction should be rounded to the whole number because it is greater than the fraction.

i) $\frac{3}{4}$ is closer to 1. That means $2\frac{3}{4} \approx 3$.

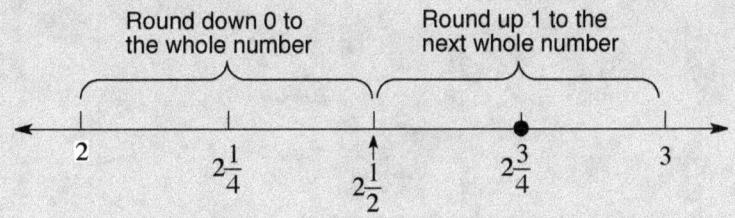

ii) $\frac{4}{9}$ is closer to 0. That means $3\frac{4}{9} \approx 3$.

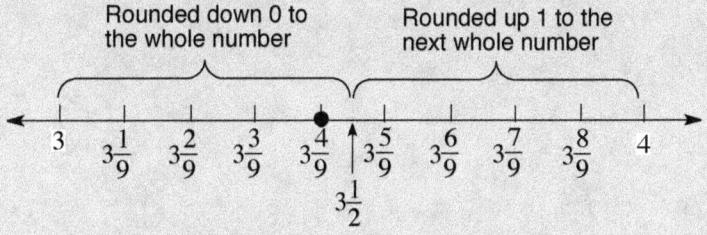

Therefore, the estimated sum of $2\frac{3}{4} + 3\frac{4}{9}$ is about 6.

Quick Exercises 9 Estimate.

1) $1\frac{3}{7} + 3\frac{7}{12}$ 2) $\frac{5}{9} + 2\frac{1}{3}$

7. Adding and Subtracting Fractions

3–19. Adding fractions.

$$2\frac{1}{5} + \frac{2}{5}$$

SOLUTION

You can add fractions with **like denominators** by adding the numerators.

i) Add numerators.

$$2\frac{1}{5} + \frac{2}{5} = (2+0)\ \overbrace{\frac{1+2}{5}} = 2\frac{3}{5}$$

iii) Let the denominators remain.

ii) Add the whole numbers.

like denominators;
the same denominators

So, the sum of $2\frac{1}{5} + \frac{2}{5}$ is $2\frac{3}{5}$.

Quick Exercises 10 Solve each expression.

1) $1\frac{1}{2} + 2$

2) $1\frac{1}{8} + 2\frac{1}{8}$

3–20. Adding fractions.

$$-\frac{1}{3} + \frac{2}{3}$$

SOLUTION

You can add fractions with **like denominators** by adding the numerators.

i) Add numerators.

$$-\frac{1}{3} + \frac{2}{3} = \overbrace{\frac{(-1)+2}{3}} = \frac{1}{3}$$

ii) Let the denominators remain.

So, the sum of $-\frac{1}{3} + \frac{2}{3}$ is $\frac{1}{3}$.

3–21. Add the fractions.

$$\frac{5}{6} + 1\frac{1}{4}$$

SOLUTION

When the fractions have mixed numbers with a whole number, first convert it to an improper fraction. If the fractions have **unlike denominators**, then find the **Least Common Denominators (LCD),**

$$\frac{5}{6} + 1\frac{1}{4} = \frac{5}{6} + \frac{5}{4}$$

i) change the mixed numbers
to improper fractions.*

ii) Rewrite the fractions as equivalent with a LCD. The LCD of $\frac{5}{6}$ and $1\frac{1}{4}$ is **12**.

iii) Now you can add fractions with **like denominators** and simplify if necessary.

$$\frac{5}{6} + \frac{5}{4} = \frac{5 \times 2}{6 \times 2} + \frac{5 \times 3}{4 \times 3} = \frac{10}{12} + \frac{15}{12} = \frac{10 + 15}{12} = \frac{25}{12} \text{ or } 2\frac{1}{12}$$

unlike denominators $\xrightarrow{\text{ii)}}$ **like denominators** $\xrightarrow{\text{iii)}}$ simplify.

So, the sum of $\frac{5}{6} + 1\frac{1}{2}$ is $2\frac{1}{12}$.

3–22. Add the fractions.

$$\left(-\frac{1}{2}\right) + \left(-1\frac{1}{3}\right)$$

SOLUTION

i) When the fractions have mixed numbers with a whole number, first convert it to an improper fraction.

$$\left(-\frac{1}{2}\right) + \left(-1\frac{1}{3}\right) = \left(-\frac{1}{2}\right) + \left(-\frac{4}{3}\right)$$

i) change the mixed numbers to improper fractions.*

ii) The LCD of $(-\frac{1}{2})$ and $(-1\frac{1}{3})$ is 6.

$$\left(-\frac{1}{2}\right) + \left(-\frac{4}{3}\right) = \left(-\frac{1 \times 3}{2 \times 3}\right) + \left(-\frac{4 \times 2}{3 \times 2}\right) = \left(-\frac{3}{6}\right) + \left(-\frac{8}{6}\right)$$

unlike denominators $\xrightarrow{\text{ii)}}$ **like denominators**

iii) When adding numbers that are negative, the sum will be always negative.

iii) Add numerators.

$$\left(-\frac{3}{6}\right) + \left(-\frac{8}{6}\right) = \frac{\overbrace{(-3) + (-8)}}{6} = \frac{-11}{6} \quad \text{or} \quad -1\frac{5}{6}$$

iv) Let the denominators remain.

So, the sum of $\left(-\frac{1}{2}\right) + \left(-1\frac{1}{3}\right)$ is $-1\frac{5}{6}$.

---

* Improper fraction: A fraction with a numerator that is greater than or equal than the denominator. For example; $\frac{3}{1}$, $\frac{6}{5}$, or $\frac{4}{4}$

Quick Exercises 11 Solve each expression.

1) $1\frac{1}{9} + 3\frac{1}{3}$

2) $2\frac{1}{2} + 1\frac{1}{5}$

3–23.   Subtracting Fractions

$$\left(-\frac{5}{7}\right) - \frac{2}{7}$$

SOLUTION

Subtracting fractions are similar to adding fractions.

Subtract numerators.

$$\left(-\frac{5}{7}\right) - \frac{2}{7} = \frac{\overbrace{(-5) - 2}}{7} = \frac{(-7)}{7} \quad \text{or} \quad -1$$

like denominators; the same denominators

Stay the same denominators.

So, the difference of $\left(-\frac{5}{7}\right) - \frac{2}{7}$ is -1.

3-24. Subtract the mixed fractions.

> **SOLUTION**
>
> If the mixed fractions have unlike denominators, then i) change the mixed numbers to improper fractions, ii) find the LCD and rewrite the fractions as equivalent, and iii) subtract and simplify if necessary.
>
> $$2\frac{1}{8} - \frac{1}{2} = \frac{17}{8} - \frac{1}{2}$$
>
> i) change the mixed numbers to improper fractions.
>
> $$\frac{17}{8} - \frac{1}{2} = \frac{17}{8} - \frac{1 \times 4}{2 \times 4} = \frac{17}{8} - \frac{4}{8} = \frac{13}{8} \text{ or } 1\frac{5}{8}$$
>
> unlike denominators $\Longrightarrow$ like denominators
>
> ii)   iii)
>
> Multiples of 8:
>   **8**, 16, 24, and so on.
> Multiples of 2:
>   2, 4, 6, **8**, and so on.
> The LCM of the denominators, 8 and 2, is 8.
> The LCD of 17/8 and 1/2 is 8.
>
> So, the difference of $2\frac{1}{8} - \frac{1}{2}$ is $1\frac{5}{8}$.

Exercises 34   Estimate the sum of the fractions and simplify if necessary.

1) $2\frac{2}{5} + \frac{5}{8}$

2) $1\frac{7}{9} + 3\frac{3}{7}$

3) $-3 + 1\frac{3}{4}$

4) $\frac{2}{3} + \frac{7}{5}$

5) $-1\frac{2}{3} + 3\frac{4}{9}$

6) $\left(-\frac{2}{5}\right) + \left(-\frac{5}{6}\right)$

Exercises 35   Add the fractions and simplify if necessary.

1) $\left(-\dfrac{2}{3}\right) + \left(-\dfrac{1}{3}\right)$

2) $\left(-\dfrac{5}{6}\right) + \left(-\dfrac{1}{6}\right)$

3) $\dfrac{6}{7} + \left(-\dfrac{2}{7}\right)$

4) $\left(-1\dfrac{1}{3}\right) + \left(-\dfrac{2}{3}\right)$

5) $2\dfrac{1}{4} + \dfrac{1}{2}$

6) $\left(-\dfrac{2}{4}\right) + \dfrac{1}{3}$

7) $\dfrac{4}{7} + \dfrac{5}{8}$

8) $\dfrac{4}{5} + \left(-\dfrac{2}{3}\right)$

Exercises 36   Solve each expression.

1)   If $x = -\dfrac{1}{2}$, find the value of $(1 + x)$.

2)   If $x = -\dfrac{2}{3}$, find the value of $(x + \dfrac{1}{2})$.

3)   If $-x = \dfrac{1}{2}$, find the value of $(\dfrac{3}{5} + x)$.

4)   If $x = -1\dfrac{1}{3}$, find the value of $(x + \dfrac{15}{18})$.

5)   If $-(x + \dfrac{1}{2}) = -5$, find the value of $(x + \dfrac{1}{2})$.

6)   If $-1 - x = \dfrac{1}{5}$, find the value of $(\dfrac{1}{5} + x)$.

7)   If $-(y + \dfrac{1}{2}) = 5$, find the value of $2(\dfrac{1}{2} + y)$.

8)   If $y - 1 = (-\dfrac{2}{3})$, find the value of $3(y - 1)$.

Exercises 37   Add the fractions and simplify if necessary.

1) $\left(-1\frac{1}{2}\right) + \frac{4}{5}$

2) $\left(-\frac{3}{8}\right) + \left(-\frac{1}{2}\right)$

3) $\left(-\frac{7}{8}\right) + 1\frac{1}{2}$

4) $\left(-\frac{5}{6}\right) + \left(-1\frac{1}{4}\right)$

5) $\left(-\frac{1}{2}\right) + \left(-\frac{1}{3}\right)$

6) $-1 + \frac{7}{9}$

7) $\frac{1}{2} + \left(-\frac{1}{3}\right)$

8) $\left(-1\frac{2}{5}\right) + \left(-\frac{1}{2}\right)$

Exercises 38   Find the value of x.

1) $x + \frac{1}{2} = \frac{1}{3}$

2) $\left(-\frac{2}{3}\right) + x = \left(-\frac{1}{3}\right)$

3) $\left(-1\frac{1}{4}\right) + x = \frac{1}{3}$

4) $x + \left(-\frac{5}{7}\right) = 1\frac{1}{3}$

5) $x + \left(-\frac{1}{6}\right) = \left(-\frac{1}{3}\right)$

6) $\left(-\frac{5}{6}\right) + x = \left(-1\frac{1}{3}\right)$

Exercises 39   Estimate the difference of the fractions and simplify if necessary.

1) $5\frac{1}{6} - 4\frac{1}{2}$

2) $\left(-1\frac{3}{4}\right) - \frac{2}{5}$

3) $1\frac{2}{3} - \left(-1\frac{3}{5}\right)$

4) $\left(-\frac{4}{9}\right) - \left(-\frac{2}{3}\right)$

5) $\left(-1\frac{3}{5}\right) - \left(-\frac{2}{3}\right)$

6) $\left(-\frac{2}{7}\right) - \left(-1\frac{1}{2}\right)$

Exercises 40   Subtract the fractions and simplify if necessary.

1) $\left(-\frac{2}{9}\right) - \left(-\frac{1}{9}\right)$

2) $\frac{1}{2} - (-1)$

3) $\left(-1\frac{1}{6}\right) - \frac{1}{6}$

4) $\left(-1\frac{1}{3}\right) - \left(-\frac{2}{3}\right)$

5) $\left(-\frac{4}{7}\right) - \left(-\frac{5}{7}\right)$

6) $\left(-\frac{2}{5}\right) - \left(-1\frac{1}{5}\right)$

Exercises 41   Solve each expression.

1) If $x = -\frac{1}{2}$ and $y = -\frac{1}{3}$, find the value of $(1 - x + y)$.

2) If $x = \frac{2}{3}$ and $y = -\frac{1}{6}$, find the value of $(x + y + 3)$.

3) If $x = \frac{1}{2}$ and $y = -\frac{2}{5}$, find the value of $(\frac{3}{5} + x - y)$.

4) If $x = -2\frac{1}{3}$ and $y = \frac{1}{2}$, find the value of $(-x \times 6) - 4y$.

Exercises 42   Subtract the fractions.

1) $\frac{1}{2} - \left(-\frac{1}{3}\right)$

2) $2\frac{1}{4} - \frac{1}{2}$

3) $\left(-\frac{3}{8}\right) - \left(-\frac{1}{2}\right)$

4) $\left(-1\frac{1}{2}\right) - \frac{4}{5}$

5) $\left(-\frac{2}{4}\right) - \frac{1}{3}$

6) $\left(-\frac{7}{8}\right) - 1\frac{1}{2}$

Exercises 43   Find the value of each variable.

1) $x - \frac{3}{4} = \frac{1}{2}$

2) $x - \left(-\frac{4}{7}\right) = \left(-\frac{2}{7}\right)$

3) $\left(-\frac{2}{3}\right) - x = \left(-\frac{1}{3}\right)$

4) $\left(-1\frac{1}{8}\right) - x = \frac{1}{4}$

5) $x - \left(-\frac{2}{3}\right) = 1\frac{1}{6}$

6) $\left(-\frac{5}{6}\right) - x = \left(-1\frac{1}{3}\right)$

Exercises 44   Solve each expression.

1)   If $x + \frac{2}{3} = -1$, find the value of $(x + \frac{2}{3})^2$.

2)   If $-2 - x = \frac{1}{3}$, find the value of $(\frac{1}{3} + x)^2$.

3)   If $-2(y + \frac{1}{2}) = 4$, find the value of $2(\frac{1}{2} + y)$.

4)   If $2(y - 1) = (-\frac{2}{3})$, find the value of $6(y - 1)$.

1. What is the difference of $\frac{3}{5} - \frac{4}{10}$?

    A. $\frac{2}{5}$ B. $\frac{1}{10}$

    C. $\frac{1}{5}$ D. $\frac{3}{10}$

2. Which of the following is the sum of $\frac{2}{7}$ and $\frac{3}{4}$?

    A. $\frac{5}{7}$ B. $\frac{5}{4}$

    C. $1\frac{1}{28}$ D. $\frac{5}{12}$

3. Jake is painting a house. It will take $5\frac{5}{12}$ hours before the house is finished. So far, it
    has been $3\frac{3}{4}$ hours. How much longer will it take for him to finish painting the house?

    A. $\frac{5}{7}$ B. $\frac{5}{4}$

    C. $1\frac{1}{28}$ D. $9\frac{1}{6}$

4. In Bill's garden, $\frac{8}{12}$ of the roses had bloomed. His wife made $\frac{1}{4}$ of the bloomed roses
    into floral arrangements. How many bloomed flowers are left?

    A. $\frac{5}{7}$ B. $\frac{5}{4}$

    C. $1\frac{1}{28}$ D. $\frac{5}{12}$

5. $\frac{1}{4}$ of students take history class while $\frac{7}{9}$ of students take science class. How many more
    students take science than history?

    A. $1\frac{1}{36}$ B. $3\frac{1}{9}$

    C. $\frac{17}{36}$ D. $\frac{9}{28}$

* Use the information for Exercises **6-7**.
  At the aquarium, the orcas are fed 50 pounds of fish. The dolphins are fed 3/5 as many pounds than the orcas.

6. How many pounds of fish do the dolphins eat?

    A. 30   **B.** 40
    C. 45   **D.** 50

7. If the seals are fed 3/4 as much as what the orcas are fed, what is the fraction showing the difference of the amount of food between the seals and the dolphins?

    A. $\dfrac{1}{5}$ **B.** $\dfrac{7}{20}$

    C. $\dfrac{3}{5}$ **D.** $\dfrac{1}{4}$

* Use the information for Exercises **8-10**.   You have \$52 in the wallet. You want to spend $\dfrac{1}{8}$ for lunch, $\dfrac{1}{10}$ for a snack, and $\dfrac{1}{5}$ for a movie.

8. What fraction shows the sum of the cost of lunch and a snack?

    A. $\dfrac{13}{40}$ **B.** $\dfrac{3}{40}$

    C. $\dfrac{1}{40}$ **D.** $\dfrac{9}{40}$

9. What fraction shows the sum of the cost of lunch and a movie?

    A. $\dfrac{13}{40}$ **B.** $\dfrac{3}{40}$

    C. $\dfrac{1}{40}$ **D.** $\dfrac{9}{40}$

10. What fraction shows the difference between the cost of a movie and lunch?

    A. \$16.65   **B.** \$14.50
    C. \$10.40   **D.** \$4.15

11. What is the sum of $2\dfrac{2}{3} + 1\dfrac{2}{6}$?

    A. $3\dfrac{4}{6}$ **B.** $4\dfrac{1}{3}$
    C. 3   **D.** 4

12. A pitcher throws the ball for $2\frac{1}{3}$ of an inning in a game. In the next game, he throws the ball for $2\frac{1}{3}$ of an inning. How many innings did he throw the ball in total?

A. $2\frac{1}{3}$

B. $4\frac{1}{3}$

C. 4

D. $4\frac{2}{3}$

13. What is the value of Δ for the equation below?
$$\frac{2}{3} - \Delta = \frac{1}{2}$$

A. $\frac{1}{3}$

B. $\frac{3}{5}$

C. $1\frac{1}{6}$

D. $\frac{1}{6}$

14. What is the value of Δ for the equation below?
$$1\frac{5}{6} - \Delta = \frac{2}{3}$$

A. $1\frac{1}{6}$

B. $1\frac{1}{3}$

C. $\frac{1}{6}$

D. $\frac{4}{3}$

15. If $-2x = \frac{3}{10}$, what is the value of the expression below?
$$4x - \frac{4}{5}$$

A. $-\frac{1}{5}$

B. $-1\frac{2}{5}$

C. $1\frac{1}{5}$

D. $\frac{1}{5}$

16. If $x + 2 = \frac{3}{5}$, what is the value of the expression below?
$$\frac{2+x}{3} + 1$$

A. -1

B. 1

C. 2

D. $1\frac{1}{5}$

8. Multiplying and Dividing Fractions.

3–25. Multiply the fractions.

$$2\frac{1}{2} \times 1\frac{1}{3}$$

SOLUTION

If they are unlike fractions, then i) change the mixed numbers to improper fractions, ii) before multiplying fractions, simplify using the GCF, iii) multiply and simplify if necessary.

$2\frac{1}{2} \times 1\frac{1}{3}$ i) Change the mixed numbers to improper fractions. $\longrightarrow$ $\frac{5}{2} \times \frac{4}{3}$

$\frac{5}{2} \times \frac{4}{3}$ ii) Before multiplying fractions, simplify with the GCF. $\longrightarrow$ $\frac{5}{\cancel{2}_1} \times \frac{\cancel{4}^2}{3}$

$$\begin{cases} \text{The factor of 2: } \mathbf{2} \times 1 = 2 \\ \text{The factor of 4: } \mathbf{2} \times 2 = 4 \\ \text{The GCF of 2 and 4 is } \mathbf{2}. \end{cases}$$

$$\frac{5}{\cancel{2}_1} \times \frac{\cancel{4}^2}{3} = \frac{5}{1} \times \frac{2}{3} = \frac{5 \times 2}{1 \times 3} = \frac{10}{3} \text{ or } 3\frac{1}{3}$$

iii) Multiply and simplify if necessary.

So, the product of $2\frac{1}{2} \times 1\frac{1}{3}$ is $3\frac{1}{3}$.

Improper fractions: A fraction with a numerator is greater than or equal than the denominator. For example; $\frac{3}{1}$, $\frac{6}{5}$, or $\frac{4}{4}$

Quick Exercises 12 Solve each expression.

1) $3\frac{1}{5} \times 1\frac{2}{8}$

2) $2\frac{2}{9} \times 4\frac{1}{2}$

3) $2\frac{1}{4} \times \frac{2}{9}$

4) $\frac{3}{5} \times \left(-\frac{2}{3}\right)$

3-26. Solve the equation.

$$\left(-\frac{1}{3}\right) \times x = \left(-1\frac{1}{6}\right)$$

SOLUTION

When the fractions are mixed numbers with a whole number, first convert it to an improper fraction.

$$\left(-\frac{1}{3}\right) \times x = \left(-1\frac{1}{6}\right) \qquad \text{Original equation}$$

$$\left(-\frac{1}{3}\right) \times x = \left(-\frac{7}{6}\right) \qquad \text{Change to improper fraction.} \quad \left(-1\frac{1}{6}\right) = \left(-\frac{7}{6}\right)$$

$$-\overset{1}{\cancel{3}} \times \left(-\frac{1}{\cancel{3}_1}\right) \times x = \left(-\frac{7}{\cancel{6}_2}\right) \times -\cancel{3}\,1 \qquad \text{Multiply } (-3) \text{ on both sides.}$$

$$x = \frac{7}{2} \text{ or } 3\frac{1}{2} \qquad \text{Simplify.} \quad (-) \times (-) = (+)$$

So, the solution of $\left(-\frac{2}{3}\right)(x) = \left(-1\frac{1}{6}\right)$ is $3\frac{1}{2}$.

Quick Exercises 13 Solve each expression.

1) $\left(-\frac{4}{5}\right) \times x = \left(-\frac{2}{5}\right)$

2) $x \times \left(-\frac{5}{8}\right) = \left(-\frac{4}{5}\right)$

3-27. Dividing fractions

SOLUTION

a) Flip the divisor.
b) Change the operation sign from division to multiplication ($\div \rightarrow \times$).
c) Multiply the fractions.
d) Simplify if necessary.

3-28. Divide the fractions.

$$3\frac{3}{5} \div 3\frac{3}{4}$$

SOLUTION

If they are unlike fractions, then i) change the mixed numbers to improper fractions, ii) flip the numerator and denominator of the divisor, iii) change the operation sign, iv) multiply and simplify if necessary.

ii) Flip (reciprocal)* of fraction.

$3\frac{3}{5} \div 3\frac{3}{4}$ — i) Change the mixed numbers to improper fractions. → $\frac{18}{5} \div \frac{15}{4} = \frac{18}{5} \times \frac{4}{15}$

iii) Change the operation sign.

$\frac{18}{5} \times \frac{4}{15}$ — iv) Multiply and simplify if necessary. →

The factor of 18: 2 x 3 x **3** = 18
The factor of 15: **3** x 5 = 15
The GCF of 18 and 15 is **3**.

$\frac{{}^{6}\cancel{18}}{5} \times \frac{4}{\cancel{15}_{5}} = \frac{24}{25}$

So, the quotient of $3\frac{3}{5} \div 3\frac{3}{4}$ is $\frac{24}{25}$.

---

* Reciprocals: two numbers are reciprocals if their product is 1. For example;

$$\frac{1}{2} \times \frac{2}{1} = 1 \qquad \frac{5}{7} \times \frac{7}{5} = 1 \qquad \frac{9}{10} \times \frac{10}{9} = 1$$

Quick Exercises 14 Solve each expression.

1) $1\frac{7}{8} \div 1\frac{1}{4}$

2) $4\frac{1}{2} \div 1\frac{2}{7}$

3–29. Solve the equation.

$$\left(-\frac{1}{2}\right) \div x = \left(-\frac{1}{4}\right)$$

SOLUTION

You can solve the problem two ways (a) and (b). The answer will be the same.
a)

$\left(-\frac{1}{2}\right) \div x = \left(-\frac{1}{4}\right)$ — i) flip(reciprocal) of fraction
ii) change the operation.
iii) then multiply them. → $\left(-\frac{1}{2}\right) \times \frac{1}{x} = \left(-\frac{1}{4}\right)$

$-\frac{1}{2x} = -\frac{1}{4}$ Simplify. $\left(-\frac{1}{2}\right) \times \frac{1}{x} = \left(-\frac{1}{2x}\right)$

$$(-2x) \times (-\frac{1}{2x}) = (-\frac{1}{4}) \times (-2x) \qquad \text{Multiply each side by } (-2x).$$

$$1 = \frac{1}{2}x \qquad \text{Simplify. } (-2x \times (-\frac{1}{2x}) = 1, (-\frac{1}{4}) \times (-2x) = \frac{x}{2}$$

$$(2) \times (1) = (\frac{1}{2}x) \times (2) \qquad \text{Multiply each side by 2.}$$

$$2 = x \qquad \text{Simplify with the GCF. } (\frac{1}{2}x \times 2) = x$$

b) When solving an equation like $-\frac{1}{2} \div x = (-\frac{1}{4})$, you can rewrite it to be $-\frac{1}{2} = (-\frac{1}{4}x)$.

$$-\frac{1}{2} \div x = (-\frac{1}{4}), \text{ or } -\frac{1}{2} = (-\frac{1}{4}x). \quad \text{Rewrite the equation.}$$

$$(-4) \times (-\frac{1}{2}) = (-\frac{1}{4}x) \times (-4) \quad \text{Multiply each side by } (-4).$$

$$2 = x \qquad \text{Simplify. } (-4) \times (-\frac{1}{2}) = 2, (-\frac{1}{4}x) \times (-4) = x$$

So, the solution of $-\frac{1}{2} \div x = (-\frac{1}{4})$, is $x = 2$.

Quick Exercises 15 Solve each expression.

1) $\quad \left(-\frac{1}{6}\right) \div x = \left(-\frac{1}{3}\right)$

2) $\quad x \div \left(-1\frac{1}{5}\right) = -1\frac{2}{3}$

Exercises 45　Multiply the fractions.

1) $\quad \frac{1}{8} \times \frac{2}{4}$

2) $\quad \frac{3}{10} \times \frac{4}{5}$

3) $\quad \frac{3}{12} \times \frac{1}{12}$

4) $\quad 2\frac{2}{3} \times \frac{1}{4}$

5) $\quad -2 \times \frac{4}{5}$

6) $\quad \frac{4}{9} \times \left(-\frac{6}{8}\right)$

Exercises 46   Solve each expression.

1) If $x = -\dfrac{3}{4}$, find the value of $x \times \dfrac{2}{3}$.

2) If $-x = 2\dfrac{1}{3}$, find the value of $\dfrac{6}{7} \times x$.

3) If $\dfrac{y}{2} = -\dfrac{1}{5}$, find the value of $(\dfrac{1}{2} \times y)$.

4) If $2(y - 1) = (-3)$, find the value of $(y - 1)$ $\times 1\dfrac{2}{3}$.

Exercises 47   Multiply the fractions.

1) $\dfrac{2}{6} \times \left(-\dfrac{3}{4}\right)$

2) $\dfrac{2}{9} \times \left(-1\dfrac{1}{5}\right)$

3) $\left(-\dfrac{1}{2}\right) \times \left(-\dfrac{1}{3}\right)$

4) $\dfrac{1}{8} \times (-1)$

5) $\left(-\dfrac{3}{5}\right) \times \left(-\dfrac{2}{3}\right)$

6) $\left(-2\dfrac{1}{2}\right) \times \dfrac{4}{5}$

7) $1\dfrac{2}{5} \times \dfrac{10}{14}$

8) $-1\dfrac{2}{3} \times \left(-\dfrac{3}{4}\right)$

9) $\left(-\dfrac{8}{9}\right) \times \left(-\dfrac{1}{2}\right)$

10) $1\dfrac{1}{5} \times 2\dfrac{1}{6}$

11) $\left(-\dfrac{5}{6}\right) \times \left(-\dfrac{1}{6}\right)$

12) $\left(-\dfrac{2}{3}\right) \times \left(-2\dfrac{2}{5}\right)$

13) $\left(-1\dfrac{2}{5}\right) \times \dfrac{2}{4}$

14) $\left(-1\dfrac{2}{3}\right) \times \left(-\dfrac{3}{5}\right)$

Exercises 48   Find the value of each variable.

1) $x \times \dfrac{4}{5} = \dfrac{1}{2}$

2) $x \times \left(-\dfrac{4}{7}\right) = 1\dfrac{1}{3}$

3) $\left(-\dfrac{3}{4}\right) \times x = \left(-1\dfrac{1}{3}\right)$

4) $\left(-1\dfrac{1}{4}\right) \times x = \dfrac{1}{5}$

5) $\left(-\dfrac{5}{8}\right) \times x = \left(-\dfrac{4}{10}\right)$

6) $x \times \left(-\dfrac{2}{3}\right) = \left(-\dfrac{4}{9}\right)$

Exercises 49   Solve each expression.

1) If $5(x + 1) = -2$, find the value of $3(x + 1) \times 1\dfrac{2}{3}$.

2) If $2 - x = \dfrac{3}{5}$, find the value of $\dfrac{2 - x}{3} \times 2\dfrac{1}{2}$.

3) If $-y + \dfrac{1}{2} = 4$, find the value of $4 \times (-4 - y)$.

4) If $(y - 1) = \left(-\dfrac{2}{3}\right)$, find the value of $\dfrac{y - 1}{2} \times \left(-\dfrac{1}{2}\right)$.

Exercises 50   Divide the fractions.

1) $\dfrac{2}{6} \div \left(-\dfrac{3}{4}\right)$

2) $-2 \div \dfrac{5}{6}$

3) $\dfrac{4}{9} \div \left(-\dfrac{4}{6}\right)$

4) $2\dfrac{1}{4} \div \dfrac{1}{2}$

Exercises 51   Solve each expression.

1) If $\dfrac{x+2}{3} = -4$, find the value of $(x+2) \div 12$).

2) If $-\left(\dfrac{1-x}{7}\right) = \dfrac{4}{5}$, find the value of $(1-x) \div 14$.

3) If $-(y+4) = \dfrac{2}{5}$, find the value of $5[(4+y) \div 2]$.

4) If $2(y-1) = \left(-2\dfrac{2}{3}\right)$, find the value of $(y-1) \div 2$.

Exercises 52   Divide the fractions.

1) $2\dfrac{2}{3} \div \dfrac{1}{4}$

2) $\left(-2\dfrac{1}{3}\right) \div \dfrac{7}{9}$

3) $\left(-\dfrac{3}{4}\right) \div \left(-\dfrac{1}{2}\right)$

4) $2\dfrac{2}{7} \div \dfrac{4}{7}$

5) $\left(-\dfrac{3}{8}\right) \div \left(-\dfrac{1}{2}\right)$

6) $\left(-\dfrac{3}{5}\right) \div \left(-1\dfrac{1}{4}\right)$

7) $-2 \div \dfrac{4}{5}$

8) $\dfrac{2}{3} \div \left(-\dfrac{1}{3}\right)$

9) $\dfrac{4}{3} \div \left(-\dfrac{8}{9}\right)$

10) $\dfrac{1}{2} \div (-1)$

11) $\dfrac{3}{12} \div \dfrac{1}{12}$

12) $2\dfrac{1}{4} \div 3$

13) $1\dfrac{1}{8} \div \dfrac{3}{2}$

14) $\left(-\dfrac{5}{7}\right) \div 5$

Exercises 53   Find the value of each variable.

1) $x \div \dfrac{1}{2} = \dfrac{2}{3}$

2) $\left(-\dfrac{2}{3}\right) \div x = \left(-\dfrac{1}{2}\right)$

3) $\left(-1\dfrac{1}{4}\right) \div x = \dfrac{5}{6}$

4) $x \div \left(-1\dfrac{4}{6}\right) = \left(-\dfrac{3}{5}\right)$

5) $x \div \left(-\dfrac{3}{4}\right) = 1\dfrac{1}{3}$

6) $\left(-\dfrac{5}{6}\right) \div x = \left(-3\dfrac{1}{3}\right)$

* Solving Problems

Exercises 54   Solve each problem using the given information.

1) Tracy has 28 movies in his collection. If 1/2 of the movies are comedies, 1/3 of the movies are action-adventure, and the rest are dramas, how many of each genre does she have? Show your work.

2) On a history test, Mike scores $\dfrac{14}{16}$ as many points as Bob, who scores 92 points. Sheena scores $1\dfrac{1}{8}$ as many points as Mike. How many points does Sheena score? Show your work and round to the nearest tenth.

3) Jane is sawing wood to make a shelf. She originally had a 15 inches board. After sawing apiece off, she now has a $9\dfrac{1}{6}$ inch board. How many inches did she saw off the board? Show your work.

SELF-TEST

1. Which equation is correct?

    A. $1\frac{5}{8} \times \frac{3}{4} = \frac{13}{8} \div \frac{4}{3}$

    B. $1\frac{5}{8} \times \frac{3}{4} = \frac{8}{13} \div \frac{4}{3}$

    C. $2 \times \frac{3}{4} = \frac{3}{4} \div \frac{2}{1}$

    D. $2 \times \frac{2}{3} = (3 + 3) \times \frac{2}{3}$

2. Which equation is correct?

    A. $2\frac{3}{5} - 1\frac{3}{9} \approx 2 - 1$

    B. $4\frac{5}{7} + 1\frac{1}{4} \approx 4 + 1$

    C. $2\frac{1}{3} \times 1\frac{5}{7} \approx 2 \times 2$

    D. $4\frac{5}{9} \div 2\frac{5}{12} \approx 4 \div 2$

3. Which equation is correct?

    A. $1\frac{1}{2} - 1\frac{4}{7} \approx 2 - 1$

    B. $1\frac{2}{3} + 1\frac{4}{5} \approx 2 + 2$

    C. $6\frac{1}{3} \times 1\frac{1}{4} \approx 6 \times 2$

    D. $2\frac{7}{15} \div 2\frac{7}{10} \approx 3 \div 3$

4. Estimate the difference of the mixed fractions.
    $$6\frac{2}{5} - 2\frac{3}{4}$$

    A. 3

    B. 4

    C. 5

    D. 6

5. Farmer Jones receives $2\frac{5}{8}$ bottles of milk every week from the milkman. Which of the following expressions correctly describes how many weeks it would take for Farmer Jones to receive 15 bottles of milk?

    A. $2\frac{5}{8} \div w = 15$

    B. $2\frac{5}{8} \times w = 15$

    C. $2\frac{5}{8} - w = 15$

    D. $2\frac{5}{8} + w = 15$

6. Leonard is spreading $1\frac{1}{4}$ pounds of fertilizer evenly over some lawns. Which of the following equations correctly describes how many lawns it would take for Leonard to spread $\frac{1}{4}$ pounds over each lawn. Use x for the number of lawns.

A. $1\frac{1}{4} \div w = \frac{1}{4}$

B. $\frac{1}{4} \times 1\frac{1}{4} = w$

C. $1\frac{1}{4} \times w = \frac{1}{4}$

D. $w + 1\frac{1}{4} = \frac{1}{4}$

7. Estimate the sum of the mixed fractions.
$$3\frac{3}{8} + 4\frac{3}{4}$$

A. 6

B. 7

C. 8

D. 9

8. Estimate the product of the mixed fractions.
$$5\frac{5}{6} \times 4\frac{3}{8}$$

A. 20

B. 24

C. 25

D. 30

9. Find the quotient.
$$2\frac{4}{7} \div 1\frac{4}{5}$$

A. 2

B. $\frac{1}{2}$

C. 4

D. $1\frac{3}{7}$

10. $\frac{5}{8}$ of the students in a class was divided into 5 groups. How many students are there in each group if there are 32 students in total?

A. 2

B. 3

C. 4

D. 6

11. Mrs. Hunter has $360.00 and she wants $\frac{4}{5}$ of the money to be divided evenly among

her four children. Given that each child receives the same amount of money, how much money did each child get?

A. $36.00   B. $288.00

C. $72.00   D. $144.00

* For Exercises **12-13**. The workers at a ranch pick apples every day. They finished one quarter of an acre yesterday and worked $\frac{7}{16}$ acres out of the remaining $\frac{3}{4}$ acres today.

12. How many acres did they work so far?

A. $\frac{7}{12}$ B. $\frac{1}{4}$

C. $\frac{10}{12}$ D. $\frac{7}{16}$

13. How many acres are left?

A. $\frac{1}{2}$ B. $\frac{1}{3}$

C. $\frac{1}{4}$ D. $\frac{1}{6}$

14. What is the value of Δ for the equation below?

$$2\frac{2}{\Delta} \times 2\frac{1}{2} = 6$$

A. 2   B. $\frac{1}{2}$

C. 5   D. $1\frac{1}{2}$

15. What is the value of Δ for the equation below?

$$2\frac{1}{4} \div \frac{3}{\Delta} = 6$$

A. 2   B. 4

C. 6   D. 8

16. What is the value of Δ for the equation below?

$$\Delta \div \frac{3}{4} = \frac{8}{9}$$

A. $\frac{2}{3}$

B. $\frac{3}{2}$

C. $\frac{9}{4}$

D. $\frac{3}{4}$

17. What is the value of Δ for the equation below?

$$\Delta \times \frac{2}{5} = \frac{2}{100}$$

A. $\frac{2}{10}$

B. $\frac{2}{100}$

C. $\frac{1}{20}$

D. $\frac{1}{40}$

18. If $-(2x + 1) = \frac{3}{4}$, what is the value of the expression below?

$$4 \times \frac{2x + 1}{3}$$

A. -1

B. $-1\frac{1}{4}$

C. $-1\frac{1}{4}$

D. -3

19. If $\frac{1 - 2x}{2} = 6$, what is the value of the expression below?

$$\frac{1 - 2x}{6} + 1$$

A. -1

B. 3

C. 4

D. $1\frac{1}{6}$

CHAPTER 4
Percentage

In this chapter, you will solve problems that involve fractions, decimals, and percentages that can apply to your daily life.

1. Understanding Percentage

A percentage is a number or ratio expressed as a fraction of 100.

4-1. What is the percentage of the shaded area below given that it is $\frac{7}{15}$ of the rectangle?

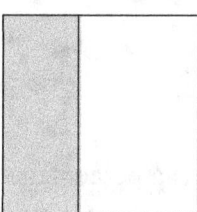

SOLUTION

Use the equation to convert the fraction into a percentage.

*Let x represent the unknown number.

$$\boxed{\text{Fraction} \times 100\% = x}$$

$\frac{7}{15} \times 100\% = x$ Substitute the equation with $\frac{7}{15}$.

$\frac{7}{\cancel{15}_3} \times \cancel{100}^{20}\% = x$ Divide 15 and 100 by 5
                                  The GCF of 15 and 100 is 5.

$\frac{140}{3}\% = x$ Multiply.

$46.7\% \approx x$ Divide 140 by 3.

So, $\frac{7}{15}$ of the rectangle is about 46.7%.

4-2. Write $1\frac{5}{12}$ as a percentage.

> **SOLUTION**
>
> First, the mixed fraction should be changed into an improper fraction.
>
> $1\frac{5}{12} = \frac{17}{12}$ Convert to an improper fraction.
>
> Let x represents the unknown percentage.
>
> Fraction $\times$ 100% = x Apply to an equation.
>
> $\frac{17}{12} \times 100\% = x$ Substitute "Fraction" with $\frac{17}{12}$.
>
> $\frac{17 \times 100\%}{12} = x$ Rearrange.
>
> $\frac{1700\%}{12} = x$ Multiply 17 by 100.
>
> $141.7\% \approx x$ Divide 1700 by 12.
>
> So, $1\frac{5}{12}$ equals to about 141.7%.

Exercises 1   Write each fraction as a percentage. Round to the nearest tenth if necessary.

1) $\frac{3}{100}$

2) $\frac{6}{200}$

3) $\frac{7}{19}$

4) $\frac{5}{7}$

5) $\frac{3}{16}$

6) $\frac{25}{28}$

7) $\frac{11}{13}$

8) $\frac{17}{36}$

9) $\frac{4}{15}$

10) $1\frac{3}{24}$

* Solving Problems

Exercises 2   Use the information for Exercises **1-4**. A pizza with 16 slices has 1260 calories.

1) If you eat 2 slices of pizza, what percentage of the pizza did you eat?

2) If you eat $\frac{1}{4}$ of the pizza, what percentage of the pizza did you eat?

3) If you eat $\frac{3}{8}$ of the pizza, what percentage of the pizza did you eat?

4) If you ate 352.8 calories, what percentage of the pizza did you eat?

Exercises 3   Use the information for Exercise **1-3**. On Monday, an ice cream store sold 57 vanilla ice cream cones, 21 strawberry ice cream cones, and 82 chocolate ice cream cones. On Tuesday, they sold 15% less ice cream then they did on Monday.

1) How many ice cream cones did they sell on Tuesday?

2) Name the percentage of vanilla ice cream cones sold on Monday.

3) If they sold 67 chocolate ice cream cones on Tuesday, what is the percentage of chocolate ice cream sold on Monday?

Exercises 4   Use the information for Exercises **1-4**. You have $462.00. You want to buy your friend a present that costs $37.50. Additionally, you are feeling hungry and want to buy a ham sandwich, which costs $6.50.

1) How much money will you have left over? What is the percentage of the money you have remaining?

2) There is a sale going on and all items are now 25% off. There is a 9% sales tax. You want to buy your friend a present that costs $35.00. How much does the present cost in total?

3) What if you decided to buy a different present, one that costs $22.00? How much does the present cost?

4) If there was a 8.5% sales tax, how much would you have to pay for a present that cost $16.50?

Exercises 5   Use the information for Exercises **1-3**. A sculptor sells sculptures that cost from $25 to $50.

1) A customer would like to buy a sculpture that costs $35. How much money does the customer have to pay if there is an 8% sales tax?

2) A customer would like to buy three sculptures that each cost $30. How much money does the customer have to pay if there is an 8% sales tax?

3) What if the sculptor offered a 25% discount for a sculpture that costs $36? A customer would like to buy two of the same discounted sculpture. How much would the customer have to pay if there is no sales tax?

Exercises 6   Use the information for Exercises **1-4**.
Lucy initially has $1,023.00 in her bank account. For 4 weeks, she deposited $35.00 weekly.

1) How much money does Lucy have in her bank account now?

2) Lucy needs to withdraw 1/3 of her money from her bank account. How much money does she have now?

3) If Lucy deposits 5/6 of what she had originally, how much money does she have now?

4) If she withdraws 35% of what she had originally, how much money does she have now?

Exercises 7   Use the information for Exercises **1-3**. A plane is depositing powdered fire retardant on a forest fire that is 250 acres long. On the first run, the plane dumps powder over 42 acres. On the second run, the plane dumps over 51 acres.

1) Name the fraction of the number of acres that was deposited on the first and second runs.

2) What is the percentage of acres that the plane dumped over in total?

3) When the plane is going on another run, it dumps over 2/3 as many acres than it did over the first and second run. What is the percentage of acres that the plane just dumped over?

Exercises 8   Use the information for Exercises **1-4**. A granola bar recipe calls for 3/8 cups of walnuts and 2/3 cups of almonds.

1) What is the percentage of cups of walnuts used in the recipe?

2) If the number of cups of walnuts are reduced by 10% in the recipe, then what is the total number of cups used?

3) If 15% more almonds are used in the recipe, how many cups of almonds are being used?

SELF-TEST

1. Which of the following percentages is equal to $\frac{15}{27}$?

   A. 15%   B. 27%
   C. 180%   D. 55.6%

2. Betsy spends 45% of her time doing her homework. What is the fraction of the time she spends NOT doing her homework?

   A. $\frac{45}{100}$ B. $\frac{18}{40}$

   C. $\frac{2}{5}$ D. $\frac{9}{20}$

3. An electronic store is announcing that all items in the store are $\frac{1}{8}$ off. What is the percentage of the discount?

   A. 8%   B. 16%
   C. 12.5% D. 25%

4. Which of the following is $\frac{5}{9}$ written as a percentage? Round to the nearest whole number.

   A. 28%   B. 56%
   C. 42%   D. 64%

5. At the parking lot, 1 out of 8 cars are white. What is the percentage of white cars?

   A. 6.25% B. 8.3%
   C. 12.5% D. 18.8%

6. David and two of his friends each blow up 15 balloons. If they need 70 balloons, what is the percentage of balloons left? Round to the nearest whole number.

   A. 36%   B. 50%
   C. 64%   D. 78%

7. Which of the following is $2\frac{2}{3}$ written as a percentage? Round to the nearest whole number.

   A. 34%   B. 68%
   C. 134%   D. 268%

8. Ben spent $2\frac{1}{2}$ hours studying on Saturday and $1\frac{3}{4}$ hours on Sunday. What is the percentage of time Ben spent on Saturday then on Sunday? Round to the nearest whole number.

   A. 40%   B. 29%
   C. 59%   D. 24%

9. Laura spent $2\frac{1}{4}$ hours in the garden. For $\frac{3}{4}$ of an hour, she planted some seeds, for an additional $\frac{1}{3}$ of an hour, she watered the plants, and then she spent another hour weeding the garden. For the rest of the time, she trimmed the garden. What is the percentage of the time she spent trimming the garden? Round to the nearest whole number.

   A. 33%   B. 15%
   C. 52%   D. 48%

2.   Change Decimals into Percentage

4–3.   Use the equation to derive a percentage from a decimal. It is done the same way as a fraction.

> Let x represents the unknown percentage.
>
> **Decimal x 100% = x**

4–4.   Rose lives 1.5 miles away from school. If Barbara's mother only drives her 1.35 miles of the way, what is the percentage of the distance she is driven?

SOLUTION

Use the equation above.

Decimal or Fraction × 100% = x Apply the equation.

$\dfrac{1.35}{1.5} \times 100\% = x$ Substitute "Decimal or Fraction" with 1.35/1.5.

$\dfrac{135}{1.5}\% = x$ Multiply 1.35 by 100.

$90\% = x$ Divide 135 by 1.5.

So her mother drives her 90% of the way.

Exercises 9   Write each decimal as a percentage.

1)	0.003	**2)**	23.48
3)	0.75	**4)**	1.02
5)	0.56	**6)**	0.07
7)	0.73	**8)**	2.01
9)	1.401	**10)**	0.2051

Exercises 10   Find the value of each expression and write as a percentage.

1) $48\% + 0.25$

2) $5\% + \dfrac{9}{12}$

3) $19.62\% - 0.11$

4) $0.061 + 0.4148\%$

5) $\dfrac{15}{120} + 15\%$

6) $0.8\% + 0.02$

7) $72\% - \dfrac{21}{100}$

8) $1.05\% + 0.05$

Exercises 11   Write each value as a percentage.

1) 0.202

2) $1\dfrac{3}{7}$

3) $2\dfrac{5}{7}$

4) 1.035

5) $\dfrac{15}{120}$

6) 0.002

7) $\dfrac{21}{18}$

8) 1.050

9) 0.014

10) $3\dfrac{5}{12}$

* Solving Problems

Exercises 12   Use the information for Exercises **1-4**.
Josh has a bouquet with 36 roses and tulips.

1) What is the percentage of roses in the bouquet if there are 5 roses?

2) What is the percentage of tulips in the bouquet if there are 5 roses?

3) If 50% of the bouquet contains tulips, how many roses are there?

4) If 25% of the bouquet contains roses, how many tulips are there?

Exercises 13   Use the information for Exercises **1-4**. The science club needs $1,000 in order to buy new equipment so they are asking for donations. Tom donated $130, Vanessa donated $165, Sammy donated $210, and Penny donated some money. Altogether, they donated $855.

1) What is the percentage of the money Tom donated?

2) If Ronald would like to donate 20% more than Tom, how much more money does he need?

3) If the sum of the donations is 78% everyone except Penny accumulated together, how much money did Penny donate?

4) If Penny additionally donated $162.00, what is the percentage of money still needed?

Exercises 14   Benji bought 68 buttons at the local craft store. Unfortunately, as he was leaving he fell and some of them fell into an open sewer. When he got home, he counted the remaining buttons and discovered that he has 23 buttons left. What is the percentage of buttons lost?

Exercises 15   Use the information for Exercises **1-4**. Leonard is spreading $10\frac{1}{4}$ pounds of fertilizer evenly over 2 differently sized lawns.

1)   If 45% of the fertilizer is used on the first lawn, how many pounds of fertilizer was used for the second lawn?

2)   If 70% of the fertilizer is used on a lawn, how many pounds of fertilizer did he use?

3)   If he is used $\frac{3}{7}$ pounds of fertilizer on a lawn, what is the percentage of fertilizer he used?

4)   If he is used $\frac{5}{6}$ pounds of fertilizer on a lawn, what is the percentage of fertilizer he used?

Exercises 16   Use the information for Exercises **1-4**. 45% of Mrs. Steven's class owns a pet. Of that number, 75% of the pets are dogs.

1)   What fraction of the class owns a pet?

2)   What fraction of the class has a pet that is not a dog?

3)   If she has 18 students in her class, then how many students own a pet?

4)   If $\frac{1}{3}$ of her class are boys, and 66.5% of them owns a pet and there are 18 people in her class, then how many boys are pet owners?

Exercises 17   Use the information for Exercises **1-2**.
Randy put 0.87 ounces of birdseed in the bird feeder. Two hours later, he found 0.64 ounces of birdseed left in the feeder.

1)   What is the percentage of birdseed eaten during the two hours Randy was gone?

2)   After a day, he found 0.46 ounces of birdseed left in the feeder. What is the percentage of birdseed eaten by the end of the day?

SELF-TEST

1. Which of the following is 0.895 written as a percentage?

    A. 10.5%   B. 21%
    C. 89.5%   D. 88%

2. Which of the following is 0.492 written as a percentage?

    A. 49.2%   B. 1.6%
    C. 100%   D. 50.8%

3. The radii of a quarter and dime are about 1.2 cm and 0.9 cm respectively. How much larger is the radius of a quarter than a dime?

    A. 9%   B. 12%
    C. 30%   D. 70%

4. It rained 0.14 inches on Monday. On Tuesday, it rained 0.11 inches less than on Monday. What is the percentage showing how much less it rained on Tuesday? Round to the nearest whole number.

    A. 14%   B. 11%
    C. 25%   D. 3%

5. AJ has a total of $46.35 in his bank account. If he withdraws $18.50 and deposits $35.20, what is the percentage showing how much money he has left in his account?

    A. 39.9%   B. 12.5%
    C. 75.9%   D. 134%

3. Solving Problems with Percentage

4–5. Complete the sentence below.
         24% of $250.00 is (   ).

SOLUTION

Use the same equation and let x represents the variable.
              24% of $25 0.00 is (   )

$$\boxed{\dfrac{1}{100\%} \text{ × Given Percent (\%) × Given Number} = x}$$

$\dfrac{1}{100\%} \times 24\% \times 250 = x$ Substitute.

You can cancel out the units (%)and simplify before multiplying

$\dfrac{1}{\cancel{100\%}_{25}} \times \overset{6}{\cancel{24\%}} \times 250 = x$ Divide 100 and 24 by the GCF 4.

Divide again until you get the lowest prime numbers.

$\dfrac{1}{\cancel{100\%}_{\cancel{25}_1}} \times \overset{6}{\cancel{24\%}} \times \overset{10}{\cancel{250}} = x$ Divide 25 and 250 by the GCF 25.

$60 = x$ Simplify.
So, 24% of $250.00 is $60.00.

Exercises 18   Solve each percentage problem.

1) 5% of $25.00 is (   ).

2) 18% of 250 is (   ).

3) 90% of 80 is (   ).

4) 110% of $25.00 is (   ).

5) 4% of $13.00 is (   ).

6) 150% of $20.00 is (   ).

7) 23% of $52.00 is (   ).

8) 0.05% of $25.00 is (   ).

9) 10% of $19.00 is (   ).

10) 18% of $74.00 is (   ).

11) 2.5% of $49.00 is (   ).

12) 36% of $28.00 is (   ).

Exercises 19   Solve each expression.

1)   20% of $38.00

2)   50% of $150.00

3)   0.2% of $30.00

4)   15% of $280.00

5)   92% of $100.00

6)   0.5% of $10.00

7)   0.1% of 100

8)   200% of $50

Exercises 20   Solve each expression.

1)   What is 0.15% of $120.00?

2)   0.55% of $15.00

3)   What is 0.02% of $100.00?

4)   What is 31% of $54.00?

5)   12% of $25.00 is (   ).

6)   What is 4% of $30.00?

7)   150% of $80.00

8)   0.05% of $180 is (   ).

9)   92% of 300 is (   ).

10)   What is 20% of $30.00?

11)   What is 30% of $80.00?

12)   3% of $500.00 is (   ).

4–6. Rose lives 1.5 miles away from school. If her mother drives her 82% of the way, how many more miles does she have to walk to get to school?

SOLUTION

You can solve this kind of problems by using multiplying the given numbers and then subtracting the value from the original distance.

* Let x represents the unknown number.

$$\frac{1}{100\%} \times \text{Given Percent (\%)} \times \text{Given Number} = x$$

Use the equation above and substitute in the known values.

$$\frac{1}{100\%} \times 82\% \times 1.5 = x \qquad \text{Substitute.}$$

You can cancel the units (%) and simplify before multiplying.

$$\frac{1}{\underset{50 \ 2\times50=100}{\cancel{100\%}}} \times \overset{41 \quad 2\times41=82}{\cancel{82\%}} \times 1.5 = x \qquad \text{Divide 100 and 82 by the GCF 2.}$$

$$\frac{61.5}{50} = x \qquad \text{Multiply 41 by 1.5.}$$

$$1.23 = x \qquad \text{Divide 61.5 by 50.}$$

Rose's mother drives her 1.23 miles of the way.
1.5 miles – 1.23 miles = 0.27 miles
So, she has to walk for 0.27 miles.

4–7. If 0.3% of x is 15, what is 20% of x?

SOLUTION

First solve for the variable then insert the value in the second expression.

* Let x represents the variable of an unknown number.

If 0.3% of x is 15, what is 20% of x?

$$\frac{1}{100\%} \times \text{Given Percent (\%)} \times x = \text{Given Number}$$

Use the equation above and substitute in the known values.

$$\frac{1}{100\%} \times 0.3\% \times x = 15 \qquad \text{Substitute.}$$

You can cancel the units (%) and simplify before multiplying.

$$\frac{0.3\%}{100\%} \times x = 15 \qquad \text{Simplify.} \quad \frac{1}{100\%} \times 0.3\% = \frac{0.3\%}{100\%}$$

$$0.003x = 15 \qquad \text{Simplify.} \quad \frac{0.3\%}{100\%} = 0.003$$

$$x = 5000 \qquad \text{Divide both sides by } 0.003.$$

After you found the variable, you can solve the problem.

* Let y represents the variable of an unknown number.

$$x = 5000$$

If 0.3% of x is 15, what is 20% of x?

$$\boxed{\frac{1}{100\%} \times \textbf{Given Percent (\%)} \times \textbf{Given Number} = y}$$

$$\frac{1}{100\%} \times 20\% \times 5000 = y \qquad \text{Substitute.}$$

$$\frac{1}{\underset{5}{\cancel{100\%}}} \times \overset{1}{\cancel{20\%}} \times \overset{1000}{\cancel{5000}} = y \qquad \begin{array}{l}\text{Simplify 100 and 20 using the GCF of 20.} \\ \text{Simplify 5 and 5000 using the GCF of 5.}\end{array}$$

$$1000 = y \qquad \text{Simplify.}$$

So, the solution is 1000.

Exercises 21   Solve each percentage problem.

1)   If 8% of x is 30, what is 15% of x?

2)   If 15% of x is 40, what is 8% of x?

3)   If 50% of b is 150, what is 15% of b?

4)   If 40% of x is 120, what is 8% of x?

5)   If 25% of x is 80, what is 0.1% of x?

6)   If 75% of y is 15, what is 30% of y?

7)   If 30% body fat is 250 pounds, what is 25% body fat?

8)   If 8% of the sales tax is $25, what is 15% of the sales tax?

Exercises 22   Solve each percentage problem.

1)   If 8% of x is 120, what is 8% of x?

2)   If 2% of x is 15, what is 250% of x?

3)   If 0.2% of b is 1,000, what is 7% of b?

4)   If 45% of x is 500, what is 125% of x?

5)   If 25% of x is 50, what is 150% of x?

6)   If 2% of y is 15, what is 110% of y?

Exercises 23   Solve each percentage problem.

1)   If 12% body fat is 180 pounds, what is 18% body fat?

2)   If 3% of the sales tax is $825.00, what is 5% of the sales tax?

3)   If 30% of the total cost is $38.00, what is 20% of the total cost?

4)   If 25% of the body muscle is 85 pounds, what is 30% of the body muscle?

5)   If 2% of the number of rainy days are 30 days, what is 4% of number of rainy days?

6)   If 9% of the sales tax is $45.00, what is 12% of the sales tax?

7)   If 12% of the total cost is $120.00, what is 15% of the total cost?

8)   If 38% of the body muscle is 85 pounds, what is 15% of the body muscle?

9)   If 38% of rainy days are 20 days, what is 20% of the rainy days?

10)   If 4% of the sales tax is $250.00, what is 3% of the sales tax?

* Solving Problems

Exercises 24   Use the information for Exercises **1-2**. Jonathan ordered a T-bone steak that costs $35.68.

    1)   If he would like to pay 16% of the bill as a tip, how much money does he need to pay?

    2)   If he would like to pay 20% of the bill for the tip, how much money does he need to pay?

Exercises 25   Use the information for Exercises **1-2**. A store is having a 20% discount on all items. Joan and her friend want to buy some items.

    1)   If she wants to buy an item for $38.00, how much money does she have to pay, given that sales tax is 8%?

    2)   If her friend wants to buy an item for $45.00, how much money does she have to pay given that sales tax is 8%?

Exercises 26   Solve each problem using the given information.

    1)   A pizza costs $21.76. If it is divided in 8 pieces, how much does each slice cost? What is 25% of the total cost? Show your work.

    2)   There are 60 apples in 5 boxes. How many apples are in a single box? What is 60% of the apples in a box? Show your work.

SELF-TEST

1. If 15% of the total cost is $48.00, what is 25% of the total cost?

    A. 1.8   B. 18
    C. 16.2   D. 80

2. Complete the sentence.
    12.5% of $150.00 is (   ).

    A. $18.75   B. $1875.00
    C. $37.50   D. $1250.00

3. If Jeremy spent 28% of his money for a dinner show that costs $39.00, what is 40% of his money?

    A. $4.37   B. $8.74
    C. $27.30   D. $55.71

4. If 2.8% of a person's savings is $19.00, what is 3.2% of his savings?

    A. $0.53   B. $16.63
    C. $21.71   D. $21,205.36

5. Monica wants to buy an item that costs $45.00. All items in the store are 40% off. How much money does she need to pay, given that there is a sales tax of 8% before the discount?

    A. $18.00   B. $30.60
    C. $72.00   D. $112.5.00

6. What is 35% of $125.00?

    A. $18.75   B. $74.00
    C. $43.75   D. $357.14

7. If a container can hold up to 85% of the water in 92 bottles, how many bottles of water equals 55% of the container?

    A. 142   B. 167
    C. 60   D. 108

8. Complete the sentence.

12% of $52.00 is (   ).

A. $3.12

B. $6.24

C. $624.00

D. $433.33

9. If 15% is the tip for a meal that costs $13.00, what is 20% of the cost of the meal?

A. $0.39

B. $86.67

C. $65.00

D. $17.33

10. Complete the sentence.

49.5% of $320.00 is (   ).

A. $15,840.00

B. $646.46

C. $6.46

D. $158.4

CHAPTER 5
Measurements and Geometry

In this chapter, you will classify the different kinds of solids, find the various surface areas of solids, such as pyramids, prisms, and spheres and also be able to find their volumes.

1. Use the Formulas

5–1. A) Use the following formulas for this chapter.

Name	Formula	
The Distance Formula	$AB = \sqrt{(x_2 - x_1)^2 + (y_2 - y_1)^2}$:	$A(x_1,\ y_1)$ and $B(x_2,\ y_2)$
Pythagorean Theorem	$c^2 = a^2 + b^2$	c = hypotenuse, b = short leg, and a = long leg
The Midpoint Formula	$m_{AB} = \left(\dfrac{x_1 + x_2}{2}, \dfrac{y_1 + y_2}{2}\right)$:	$A(x_1,\ y_1)$ and $B(x_2,\ y_2)$
Square	$P = 4s$ $A = s^2$	P = perimeter and A = area s = side length
Rectangle	$P = 2l + 2w$ $A = lw$	P = perimeter, A = area l = length and w = width
Triangle	$P = a + b + c$ $A = (1/2)bh$	P = perimeter, A = area b = base, a and c = side lengths h = height
Circle	$C = 2\pi r$ $A = \pi r^2$	C = circumference, A = area r = radius of a circle

B) Use the following formulas for this chapter.

Name of Solid	Lateral Area (L.A.)	Surface Area (S.A.)	Volume (V)
Right Prism	L.A. = ph	S.A. = L.A. + 2B S.A. = $ph + 2(lw)$	V = Bh B = lw V = lwh
Cylinder	L.A. = Ch or L.A. = $2\pi rh$ $C = 2\pi r$	S.A. = L.A. + 2B B = πr^2 S.A. = $2\pi rh + 2\pi r^2$	V = Bh B = πr^2 V = $\pi r^2 h$
Key	p = perimeter h = height	C = circumference, h = height, r = radius	B = area of a base l = length w = width

2. Circles

5–2. Circles

i) A circle is a round plane figure whose boundary (circumference) consists of points that are equidistant from a fixed point (center).
ii) A radius is a straight line from the center to a point on the circle's circumference.
iii) The diameter is a straight line passing through the center and whose endpoints lie on the circumference.
iv) A chord is a segment whose endpoints are on a circle's circumference.

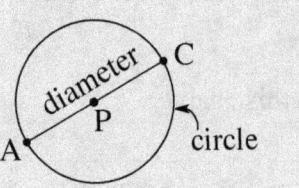

- ○ Circle: at P
- ○ **Diameter** (*d*): Line AC or $\overline{AC}$
- ○ **Radius** (*r*): Line EP or $\overline{EP}$

- ○ Center: Point P
- ○ Chord: Line FG or $\overline{FG}$

5–3. Finding the circumference and area of a circle.

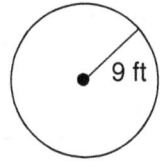

The formula for the circumference of a circle is $C = 2\pi r$ (*r* = radius)

$C = 2(3.14)r$
$C = 2(3.14)(9)$ Substitute radius (*r*) with 9
$C = 56.52$ ft   Multiply.

The formula for the area of a circle is $A = \pi r^2$ (*r* = radius.

$A = (3.14)r^2$
$A = (3.14)(9)^2$ Substitute radius (*r*) with 9.
$A = 254.34$ ft^2 Multiply.

* Perimeter: The sum of the side lengths of a polygon.

Exercises 1   Use the circle at the right. Classify each part below.

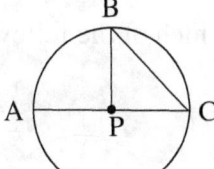

1) $\overline{AP}$ 4) $\overline{BP}$

2) $\overline{AC}$ 5) $\overline{PC}$

3) $\overline{BC}$ 6) P

Exercises 2   For Exercises 7-9, find the circumference of each circle (use $\pi = 3.14$).

1) 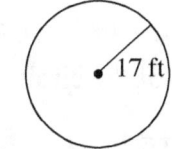   2) 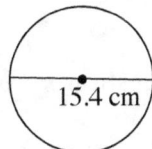   3)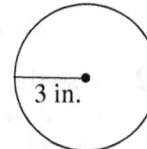

4) 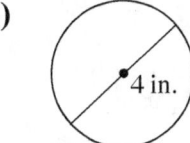   5) 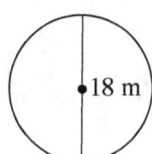   6)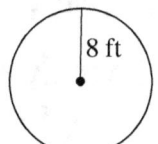

Exercises 3   Solve each problem using the given information (use $\pi = 3.14$).

1) A hula-hoop has a diameter of 1.6 ft. Find the circumference.

2) If the circumference of a circle is 15.6π cm, find the radius of the circle.

3) Find the perimeter of the square and circumference of the circle below.

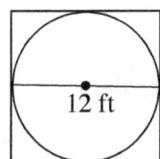

1. Which of the following is the chord of the circle below?

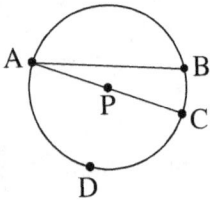

   A. $\overline{AB}$ B. $\overline{AP}$
   C. $\overline{AC}$ D. P

2. Use the diagram from Question 1. What is the radius of the circle?

   A. $\overline{AB}$ B. $\overline{AP}$
   C. $\overline{AC}$ D. P

3. If the radius of a circle is 5 in, which of the following is the circumference of the circle?

   A. 10 in.   B. 15.7 in.
   C. 78.5 in.   D. 31.4 in.

4. The diameter of a circle is 17 ft and the side length of the square is 12 ft. What is the circumference of the circle?

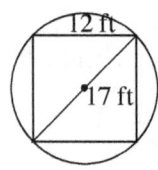

   A. 17 ft   B. 26.69 ft
   C. 53.38 ft   D. 229.9 ft

5. Use the diagram from Question 4. What is the perimeter of the square?

   A. 3 ft   B. 4 ft
   C. 48 ft   D. 68 ft

6. If the diameter of a circle is 1.2 inches, what is its circumference?

   A. 3.768 inches   B. 1.13 inches
   C. 7.536 inches   D. 2.4 inches

7. If the circumference of a circle is 16π cm, what is its radius?

    A. 16 cm   **B.** 4 cm
    C. 32 cm   **D.** 8 cm

8. Which of the following is the circumference of a circle with a diameter of 6 cm? Round to the nearest whole number.

    A. 3π **B.** 4π
    C. 5π **D.** 6π

9. Which of the following is NOT a true statement?

    A. The diameter is a line that goes across the circle, through the center.
    B. The radius is a line that passes through the center to a point on the circle.
    C. The distance of the chord is always equal to the radius.
    D. The chord is a segment whose endpoints are on a circle.

10. Given that the diameter of a circle is 4 in, what is the area of the circle? Put your answer in terms of π.

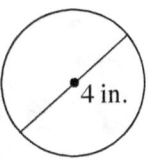

4 in.

    A. 3π in^2 **B.** 4π in^2
    C. 5π in^2 **D.** 6π in^2

11. The diameter of a coin is 2.4 cm. What is the area of the coin?

    A. 1.2π cm^2 **B.** 2.4π cm^2
    C. 1.44π cm^2 **D.** 5.76π cm^2

12. If a radius of a circle is 9 cm, what is the area of the circle?

    A. 3π cm^2 **B.** 9 cm^2
    C. 81π cm^2 **D.** 18 cm^2

13. The circumference of a circle is 14π m. What is the area of the circle? Put your answer in terms of π.

    A. 7π m^2 **B.** 49π m^2
    C. 28π m^2 **D.** 14π m^2

3. Areas and Perimeters of the Figures

5–4. Area (A) of a polygon

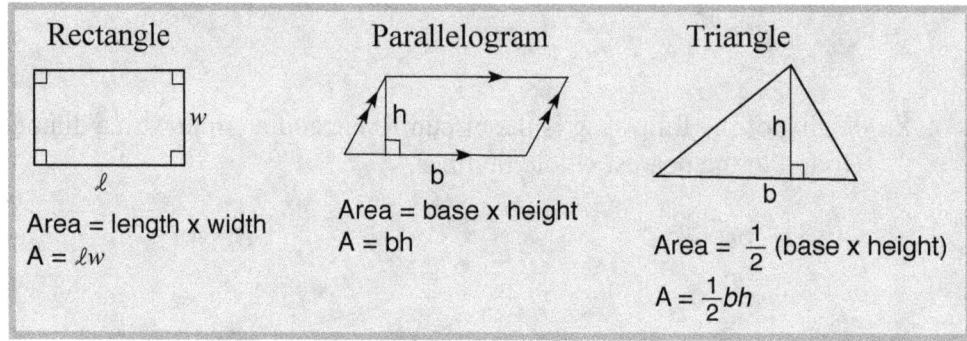

5–5. Finding the perimeter and area of a square.

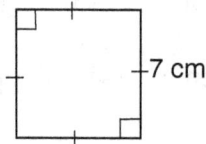

7 cm

SOLUTION

The formula for the perimeter of a square is P = 4*s* (*s* = side).

P = 4(7)   Substitute *s* with 7.

P = 28 cm   Multiply.

The formula for the area of a square is A = *s*² (*s* = side).

A = *s*²

A = 7²   Substitute *s* with 7.

A = 121 cm²   Multiply.

5–6. Finding the perimeter and area of a triangle.

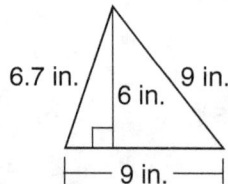

6.7 in.   6 in.   9 in.

9 in.

SOLUTION

The formula for the perimeter of a triangle is P = *a* + *b* + *c* (*a* = side length, *c* = side length, and *b* = base length).

$$P = 6.7 + 9 + 9 \qquad \text{Substitute in the values.}$$
$$P = 24.7 \text{ in.} \qquad \text{Add.}$$

The formula for the area of a triangle is $A = \frac{1}{2}bh$ (b = base, h = height).

$$A = \frac{1}{2}bh$$
$$A = \frac{1}{2}6(9) \qquad \text{Substitute base (b) with 9 and height (h) with 6.}$$
$$A = 27 \text{ in.}^2 \qquad \text{Multiply.}$$

Exercises 4   Find the area and perimeter of each figure.

1)

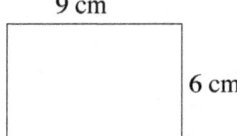

9 cm

6 cm

2)

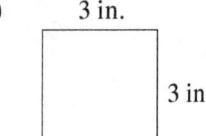

3 in.

3 in.

3)

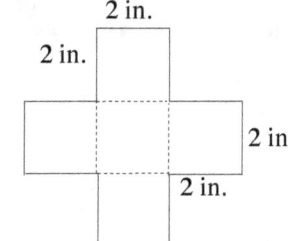

2 in.

2 in.

2 in.

2 in.

4)

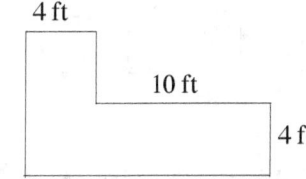

4 ft

10 ft

4 ft

5)

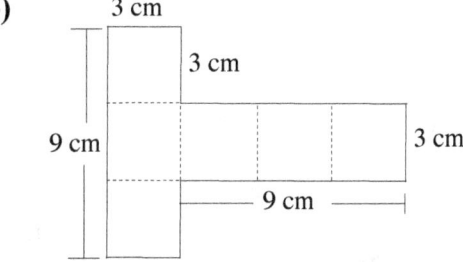

3 cm

3 cm

9 cm

3 cm

9 cm

6)

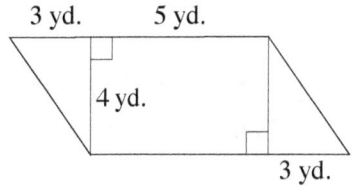

3 yd.   5 yd.

4 yd.

3 yd.

Exercises 5   Solve each problem using the given information.

1) If the area of a parallelogram is 165 m² and its height is 11 m, what is the side length of the parallelogram?

2) If the area of a rectangle is 46.75 ft² and its length is 8.5 ft, what is the width of the rectangle?

3) If the area of a triangle is 48 cm² and its height is 12 cm, what is the base length of the triangle?

SELF-TEST

1. A square has an area of 64 in.². What is the perimeter of the square?

A = 64 in.²

  A. 84 in²   **B.** 19 in
  C. 32 in   **D.** 168 in²

2. What is the area of the figure below?

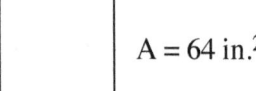

  A. 21 cm²   **B.** 21 cm
  C. 42 cm   **D.** 42 cm²

3. What is the area of the parallelogram below?

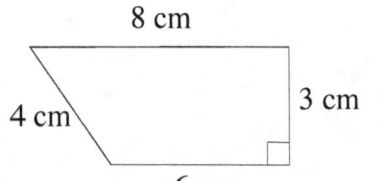

  A. 36 cm²   **B.** 18 cm
  C. 36 cm   **D.** 18 cm²

4. A triangle has an area of 24 ft^2 and a height of 8 ft. What is the perimeter of the trapezoid?

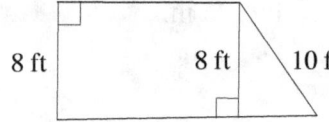

Area of triangle = 24 ft^2
Area of rectangle = 80 ft^2

    A. 24 ft   **B.** 24 ft^2
    C. 44 ft   **D.** 44 ft^2

5. What is the area of the triangle below?

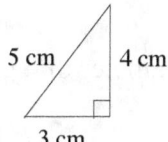

    A. 6 cm^2 **B.** 15 cm^2
    C. 20 cm^2 **D.** 12 cm^2

6. If a triangle has an area of 48 cm^2 and a base of 12 cm, what is its height?

    A. 4 cm   **B.** 8 cm
    C. 10 cm   **D.** 12 cm

7. If a rectangle has a width of 9 ft and an area of 126 ft^2, what is its length?

    A. 7 ft   **B.** 14 ft
    C. 28 ft   **D.** 23 ft

8. If the base of a right triangle is 12 ft and the height is 5 ft., which of the following is the perimeter of the triangle?

    A. 18 ft   **B.** 28 ft
    C. 30 ft   **D.** 72 ft

9. If a triangle has side lengths of 9 ft., 10ft., and 5 ft., which of the following is the perimeter of the triangle?

    A. 12 ft   **B.** 27 ft
    C. 24 ft   **D.** 29 ft

10. Find the circumference of a circle given that the radius is 4 in. Round to the nearest whole number.

    A. 50 in.   B. 25 in.
    C. 13 in.   D. 8 in.

11. Which of the following equations could be used to find the area of a rectangle with a length of 12 cm and a width of 7 cm?

    A. $A = 2(12 \times 7)$ B. $A = 12 \times 7$
    C. $A = 2(12 + 7)$ D. $A = (12 + 7)$

12. Which of the following is the area of a circle with a diameter of 6 cm? Round to the nearest whole number.

    A. 19 cm^2 B. 28 cm^2
    C. 113 cm^2 D. 226 cm^2

13. If a circumference of a circle is 20 cm, which of the following equations could be used to find the radius of the circle?

    A. $r = \dfrac{20\,\pi}{2}$ B. $r = \dfrac{2}{20\pi}$

    C. $r = \dfrac{2\,\pi}{20}$ D. $r = \dfrac{20}{2\pi}$

14. If the area of a triangle is 28 m^2 with a base length of 14 m, which of the following equation should be used to find the height of the triangle?

    A. $h = \dfrac{2(14)}{28}$ B. $h = \dfrac{14}{2(28)}$

    C. $h = \dfrac{2(28)}{14}$ D. $h = \dfrac{28}{2(14)}$

15. Which of the following are the perimeter and area of a rectangle with a length of 10 ft and a width of 5 ft?

    A. $P = 50 \text{ ft}, A = 30 \text{ ft}^2$ B. $P = 30 \text{ ft}, A = 50 \text{ ft}^2$
    C. $P = 15 \text{ ft}, A = 35 \text{ ft}^2$ D. $P = 30 \text{ ft}, A = 100 \text{ ft}^2$

16. Which of the following are the perimeter and area of a triangle given that the side lengths are 2 cm, 3 cm, and 4 cm, the base length is 4 cm and the height is 2 cm long?

A. $P = 11$ cm, $A = 12$ cm^2

B. $P = 9$ cm, $A = 8$ cm^2

C. $P = 9$ cm, $A = 4$ cm^2

D. $P = 12$ cm, $A = 4$ cm^2

17. Find the perimeter of the rectangle below.

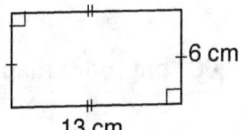

6 cm

13 cm

A. 19 cm

B. 29 cm

C. 38 cm

D. 78 cm

18. What is the area of the circle? Round to the nearest whole number.

17

A. 53 units^2

B. 107 units^2

C. 227 units^2

D. 907 units^2

4. Identifying Angles
5–7. Angles

---

1) Complementary angles: Two angles that add up to 90°.

2) Supplementary angles: Two angles that add up to 180°.

3) Vertical angles: Two nonadjacent angles that share the same vertex and are congruent.

4) Acute angles: Angles that measures less than 90° but more than 0°

5) Obtuse angles: Angles that measures more than 90° but less than 180°

5–8. In the diagram below, find an acute angle, an obtuse angle, complementary angles, supplementary angles, and congruent angles.

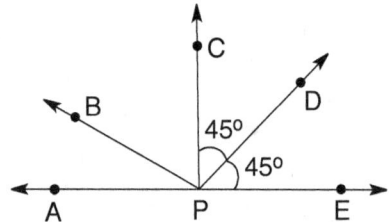

SOLUTION

1) The measure of ∠DPE is less than 90°. So, ∠DPE is an acute angle.

2) The measure of ∠BPE is greater than 90°. So, ∠AOB is an obtuse angle.

3) The sum of the measures of ∠DPE and ∠CPD is 90°. So, they are complementary angles.

4) The sum of the measures of ∠APD and ∠DPE is 180°. So, they are supplementary angles.

5) The measures of ∠DPE and ∠CPD are the same. So, they are congruent angles.

Exercises 6   Solve each problem using the given information.

1) Given that $m\angle APB = m\angle BPC$, what is the relationship between the two angles?

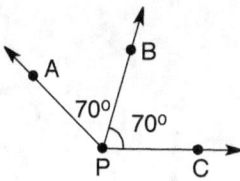

2) Find the measure of $m\angle HGI$.

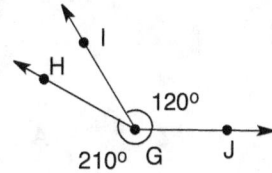

3) Name the complementary angles in the figure below.

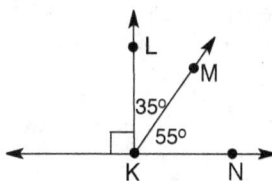

4) What is the measure of $\angle BAD$?

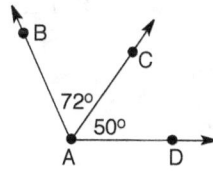

5) Suppose that the $\angle A$ and $\angle B$ are complementary angles. The measure of $\angle A$ is $56°$. What is the measure of $\angle B$?

SELF-TEST

1. The measure of ∠SPT is 70°. What is the measure of ∠SPQ?

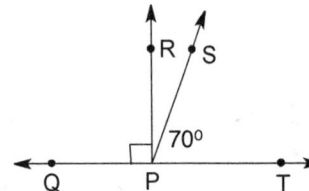

    A. 70°   **B.** 90°
    C. 110°   **D.** 20°

2. The measure of ∠ADB is 74°. What is the measure of ∠BDC?

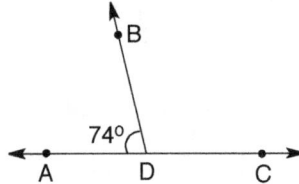

    A. 74°   **B.** 106°
    C. 180°   **D.** 16°

3. What is the measure of ∠APC?

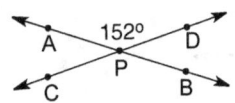

    A. 28°   **B.** 38°
    C. 48°   **D.** 58°

4. Suppose that ∠B and ∠C are supplementary angles. The measure of ∠B is 18°.
    Which equation could be used to find ∠C?

    A. $\angle C + 18° = 180°$ **B.** $\angle C - 180° = 18°$
    C. $\angle C + 18° = 90°$ **D.** $\angle C + 90° = 18°$

5. Suppose that ∠A and ∠B are complementary angles while ∠B and ∠C are
    supplementary angles. The measure of ∠C is 117°. Which equation could be used to
    find the measure of ∠A?

    A. $m\angle A + 63° = 180°$ **B.** $m\angle A - 63° = 180°$
    C. $m\angle A + 90° = \angle C$ **D.** $m\angle A - 63° = 90°$

6. Suppose that ∠A and ∠B are complementary angles while ∠B and ∠C are supplementary angles. The measure of ∠B is 26°. What is the sum of the measures of ∠A and ∠C?

    A. 64°   **B.** 154°
    C. 218°   **D.** 180°

7. If D is in the interior of ∠BAC, then ∠BAD = ∠DAC. Which of the following terms best defines this statement?

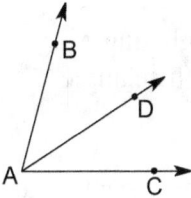

    A. Supplementary angles
    B. Complementary angles
    C. Congruent angles
    D. Ray

8. Which of the following terms best classifies ∠BAC if its measure is 135°?

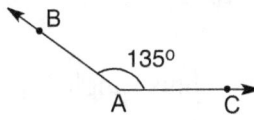

    A. Straight   **B.** Right angle
    C. Acute angle   **D.** Obtuse angle

9. Which of the following terms best classifies ∠QPR?

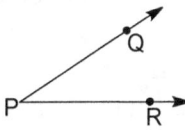

    A. Straight   **B.** Right angle
    C. Acute angle   **D.** Obtuse angle

10. Which of the following terms best classifies ∠QOT?

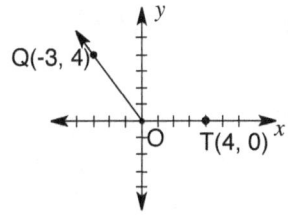

    A. Straight   **B.** Right angle
    C. Acute angle   **D.** Obtuse angle

11. Which of the following terms best classifies ∠TOS?

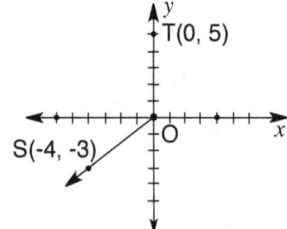

A. Straight   B. Right angle
C. Acute angle   D. Obtuse angle

12. Which of the following terms best classifies ∠ABC if its measure is 13°?

A. Straight   B. Right angle
C. Acute angle   D. Obtuse angle

13. Which of the following statements best describes a supplementary angle?

A. An angle whose measure is exactly 90°.
B. An angle whose measure that is greater than 0°, but less than 90°.
C. One of two angles of which the sum of their measures is 180°.
D. An angle whose measure is greater than 90°, but less than 180°.

14. Which of the following statements best describes an obtuse angle?

A. An angle whose measure is exactly 90°.
B. An angle whose measure that is greater than 0°, but less than 90°.
C. One of two angles of which the sum of their measures is 180°.
D. An angle whose measure is greater than 90°, but less than 180°.

15. Which of the following terms describes the angle pair ∠1 and ∠2 as shown in the diagram below?

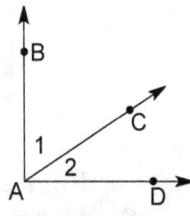

A. Supplementary angles
B. Complementary angles
C. Angle bisector
D. Perpendicular bisector

16. Name what kind of angles ∠3 and ∠4 are as shown in the diagram below.

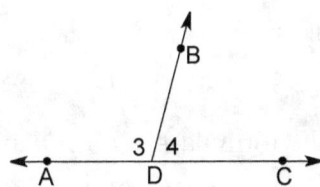

    A. Supplementary angle
    B. Complementary angle
    C. Angle bisector
    D. Perpendicular bisector

17. Which of the following statements best describes complementary angles?

    A. Two adjacent angles whose nonadjacent sides are opposite rays.
    B. Two nonadjacent angles that are vertical from each other.
    C. Two angles whose measures have a sum of $180°$.
    D. Two angles whose measures have a sum of $90°$.

5. Triangles

5–9. Triangles

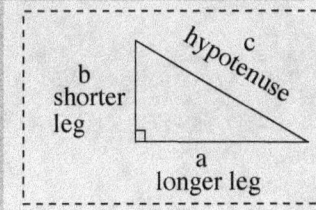

The hypotenuse (c) is the longest side of a right triangle and is the side opposite the right angle. The formula of the Pythagorean Theorem is $c^2 = a^2 + b^2$

5–10. Use the given information in the diagram to find the value of x and the area of the triangle.

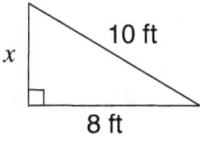

> **SOLUTION**

The Pythagorean Theorem describes the relationship between the length of the hypotenuse (c) of a right triangle and the lengths of the other two legs (a and b).

$c^2 = a^2 + b^2$ Pythagorean Theorem
$10^2 = x^2 + 8^2$ Substitute.
$100 = x^2 + 64$ Multiply.

$$100 - 64 = x^2 + 64 - 64 \qquad \text{Subtract 64 from both sides.}$$
$$36 = x^2 \qquad\qquad\qquad\quad \text{Simplify.}$$
$$6 = x \qquad\qquad\qquad\qquad \text{Square root.}$$

So, x is 6 ft.

The area of a triangle can be found using the formula A = (1/2)bh, where b is the base of the triangle and h is the height.

So A = (1/2)bh Formula of the area of a triangle

A = (1/2)(8)(6)   Substitute 8 with base length (b) and and 6 with height (h).

A = 24 ft^2 Multiply.

So, the area of the triangle is 24 ft^2

Exercises 7   In the diagram of the square and a rectangle below, the lengths of each diagonal (d) are 54 cm. What are the values of x and the perimeters of each figure?

a)

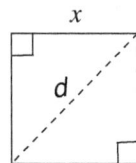

b)

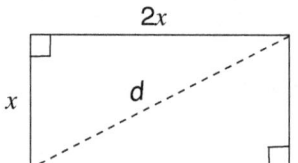

5–11.   ΔABC is similar to ΔDEF. Find the length of BC.

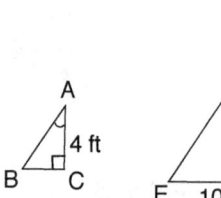

SOLUTION

$$\frac{\overline{AC}}{\overline{DF}} = \frac{\overline{BC}}{\overline{EF}} \qquad \text{Write as a proportion}$$

$$\frac{4}{16} = \frac{\overline{BC}}{10} \qquad \text{Substitute in the known values}$$

$$4(10) = (\overline{BC})(16) \qquad \text{Cross Product Property}$$
$$40 = \overline{BC}(16) \qquad \text{Multiply.}$$
$$2.5 = \overline{BC} \qquad \text{Divide each side by 16.}$$

Exercises 8 Find the length of $\overline{AC}$ in each triangle.

1)

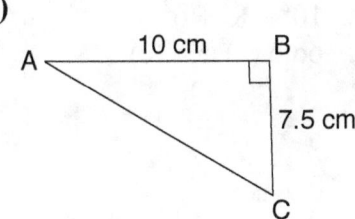

2)

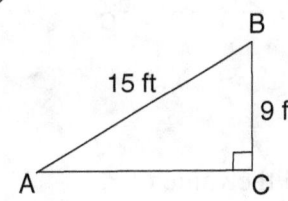

3)

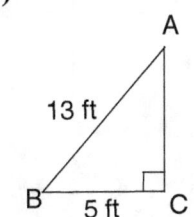

4)

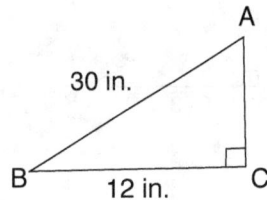

5)

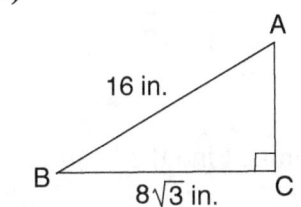

6)

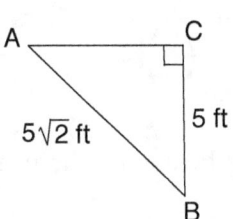

SELF-TEST

1. Use the Pythagorean Theorem to find the length of a leg of a given triangle if the length of the other leg is 25 ft and the length of the hypotenuse is 34 ft. Round your answer to the nearest tenth.

   A. 9.0   **B.** 18.0

   C. 23.0   **D.** 29.2

2. Which of the following equations is true for the right triangle below?

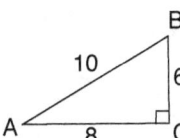

A. $8^2 = 10^2 - 6^2$
B. $10^2 = \sqrt{(8)(6)}$
C. $10^2 = 8^2 + 6^2$
D. both (a) and (c)

3. Find the value of x.

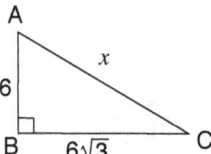

A. 8
B. $12\sqrt{3}$
C. 12
D. 10

4. Which of the following is the third length of a right triangle with two side lengths of 12 and 20?

    A. 15.5   B. 16

    C. 23.3   D. 17

5. The side length of a given square is $10\sqrt{2}$ units. Find the length of diagonal a.

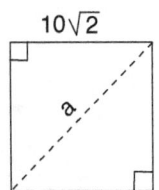

    A. 20   B. 21

    C. 12   D. 15

6. In a rectangle, the length of the diagonal is 16m. Find the value of A.

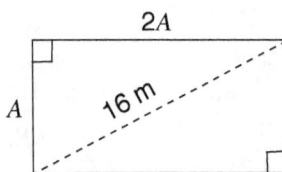

    A. 11.31   B. 9.24

    C. 7.16   D. 14.31

7. In $\triangle ABC$, $\overline{AB} = \overline{BC}$, and $\overline{AC} = 8$ yd. What is the value of x?

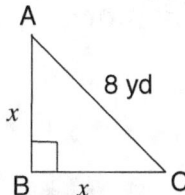

A. 32   B. $2\sqrt{2}$

C. $4\sqrt{2}$ D. $8\sqrt{2}$

8. In $\triangle ABC$, $\overline{AB} = 32$ cm, $\overline{AC}$ is twice as long than $\overline{BC}$. What is the value of x?

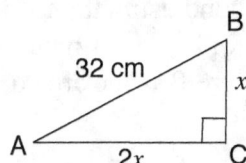

A. 18.48   B. 13.06

C. 22.63   D. 14.31

9. $\triangle ABC$ is similar to $\triangle DEF$. The ratios of the perimeter of $\triangle ABC$ to $\triangle DEF$ are 3:1. Find the perimeter of $\triangle DEF$ if the perimeter of $\triangle ABC$ is 96 in. Round your answer to the nearest tenth.

A. 288 in.   B. 48 in.

C. 32 in.   D. 96 in.

10. $\triangle ABC$ is similar to $\triangle DEF$. Find the length of DE.

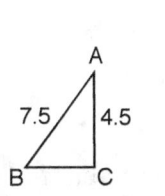 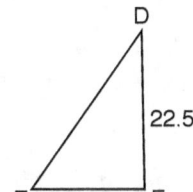

A. 28.5   B. 37.5

C. 32.5   D. 39.5

11. If the lengths of $\triangle ABC$ are 6 ft, 8 ft, and 10 ft and the ratio of the side lengths of $\triangle ABC$ to $\triangle DEF$ is 1 : 4, which of the following is the perimeter of $\triangle DEF$?

A. 96 ft   B. 48 ft

C. 24 ft   D. 130 ft

6. Surface Area of the Rectangular Prism and Cylinder

5–12. Find the surface area of a prism.

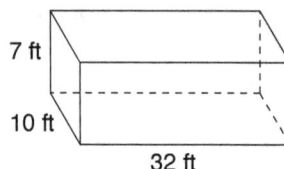

7 ft

10 ft

32 ft

SOLUTION

The surface area (S.A.) of a right prism can be found using the formula S.A. = $2lw + ph$, where l is the length, w is the width, p is the perimeter of the base and h is the height. So S.A. = 2B + L.A. where B is the area of the base and L.A. is the lateral area (ph).

$$S.A. = 2lw + ph, \text{ where } p = 2l + 2w.$$
$$= 2(10)(32) + [2(32 \text{ ft}) + 2(10 \text{ ft})](7\text{ft}) \quad \text{Substitute.}$$
$$= 640 \text{ ft}^2 + 588 \text{ ft}^2 \quad \text{Simplify.}$$
$$= 1{,}228 \text{ ft}^2 \quad \text{Add.}$$

So the lateral area is 588 ft^2 and the surface area is 1,228 ft^2.

5–13. Find the surface area of the cylinder.

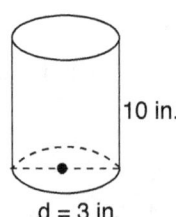

10 in.

d = 3 in.

SOLUTION

The lateral area of a cylinder is the product of the circumference of the base and height. L.A. = Ch, where C is the circumference of the base and h is the height. $C = 2\pi r$, where r is the radius of a base. The surface area of a cylinder is the sum of the lateral areas and the area of the two bases. S.A. = 2B + L.A., where B is the area of the base and L.A. is the lateral area

So L.A. = Ch, where C is the circumference of a base and h is the height. The circumference of the base of a cylinder is $2\pi r^2$.

$$L.A. = Ch = 2\pi rh$$
$$L.A. = 2\pi (1.5)(10) \quad \text{Substitute 1.5 and 10 with } r \text{ and } h.$$
$$= 30\pi \text{ in}^2 \quad \text{Multiply.}$$
$$S.A. = 2B + L.A = 2\pi r^2 + 2\pi rh$$
$$= 2\pi(1.5)^2 + 30\pi \quad \text{Substitute.}$$
$$= 34.5\pi \text{ in}^2 \quad \text{Simplify.}$$

So the lateral area is 30π in^2 and the surface area is 34.5π in^2.

7. Volume of the Rectangular Prism and Cylinder

5–14. Find the volume of the rectangular prism.

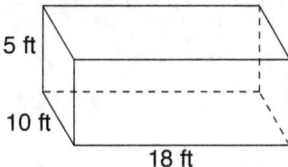

SOLUTION

The volume (V) of a prism is V = Bh, where B is the area of a base and h is the height.
The area of the base of a prism is B =lw, where l is the length and w is the width.
So, V = Bh = (lw)(h)
    = (18)(10)(5) = 900 ft^3
Therefore, the volume is 900 ft^3.

5–15. Find the volume of the cylinder. Leave your answer in terms of π.

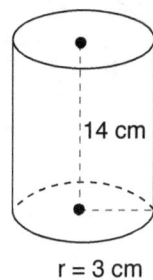

14 cm

r = 3 cm

SOLUTION

The volume (V) of a prism is V = Bh = $\pi r^2 h$, where B is the area of the base, h is the height, and r is the radius of the base. The area of the base is B = πr^2 for a circle.
So, V = Bh = $\pi r^2 h$
    = (π)(3^2)(14)
    = 126π cm^3
Therefore, the volume is 126π cm^3.

Exercises 9   Find the surface area of each figure.

1)
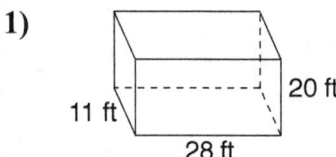
11 ft   20 ft
28 ft

2)
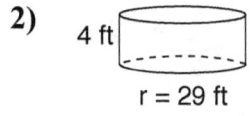
4 ft
r = 29 ft

3)
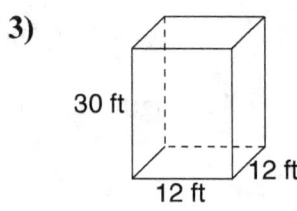
30 ft
12 ft
12 ft

4)
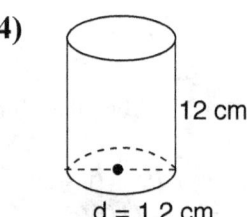
12 cm
d = 1.2 cm

Exercises 10   Find the volume of each figure.

1)

6 ft
r = 15 ft

2)
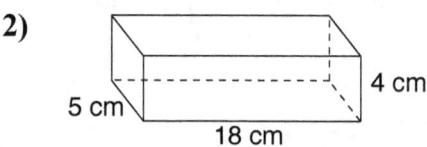
5 cm   4 cm
18 cm

3)
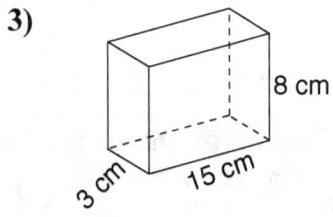
8 cm
3 cm   15 cm

4)
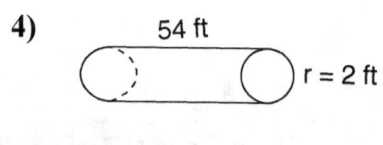
54 ft
r = 2 ft

SELF-TEST

1. If a regular rectangular prism has a base of 8 cm with a width of 5 cm, find the area of the base of the prism. Round your answer to the nearest tenth if necessary.

    A. 48.0 cm^2 **B.** 24.0 cm^2
    C. 12.0 cm^2 **D.** $12\sqrt{3} \text{ cm}^2$

2. Given that a rectangular prism has a lateral area of 224 ft^2 and its perimeter is 14 ft, find the height of the rectangular prism. Round your answer to the nearest tenth if necessary.

    A. 13 ft   **B.** 14 ft
    C. 15 ft   **D.** 16 ft

3. Which of the following statements are false?

    A.   The lateral area of a cylinder is the product of the circumference and height.
    B.   The surface area of a cylinder is the sum of twice the area of the base and the product of the base perimeter and height.
    C.   The surface area of a cylinder is S.A. $= 2\pi r^2 + 2\pi rh$.
    D.   The base area of a cylinder is $2\pi r^2$.

4. Which of the following is the area of the base of a cylinder?

    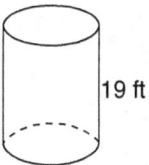
    19 ft
    r = 18 ft

    A. 2147.8 ft^2 **B.** 1073.9 ft^2
    C. 1017.4 ft^2 **D.** 2034.7 ft^2

5. A cylinder has a height of 34 m and a diameter of 5 cm. Find the area of the base of a cylinder. Round to the nearest tenth if necessary.

    A. 573.1 cm^2 **B.** 533.8 cm^2
    C. 39.3 cm^2 **D.** 157.0 cm^2

6. Find the lateral area of the rectangular prism. Round your answer to the nearest tenth if necessary.

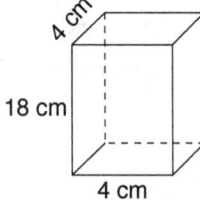

   A. 16.0 cm^2 **B.** 72.0 cm^2
   C. 144.0 cm^2 **D.** 288.0 cm^2

7. The lateral area of a cylinder is 270π in.^{2} and its height is 15 in. Find the radius of the cylinder. Round to the nearest tenth.

   A. 9 in.   **B.** 18 in.
   C. 30 in.   **D.** 1.2 in.

8. If a cylinder's height is 22 cm and the diameter of the base is 6 cm, find the surface area of the cylinder. Round to the nearest tenth if necessary.

   A. 28.3 cm^2 **B.** 445.9 cm^2
   C. 471 cm^2 **D.** 857.3 cm^2

9. The circumference of the base of a cylinder is 36π m^2 and has a height of 18 m. What is the lateral area of the cylinder? Round to the nearest tenth if necessary.

   A. 1196π m^2 **B.** 1296π m^2
   C. 972π m^2 **D.** 648π m^2

10. A cylinder has a radius of 7 ft and a height of 13 ft tall. What is the surface area of the cylinder?

   A. 140π ft^2 **B.** 280π ft^2
   C. 232π ft^2 **D.** 189π ft^2

11. Which of the following is the volume of the rectangular prism?

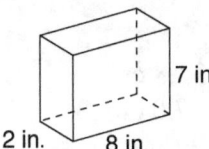

   A. 112.0 in.^{3} **B.** 56.0 in.^{3}
   C. 224.0 in.^{3} **D.** 448.0 in.^{3}

12. Find the volume of the rectangular prism.

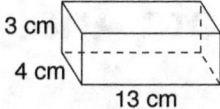

A. 78 cm^3 B. 156 cm^3
C. 312 cm^3 D. 624 cm^3

13. What is the volume of a rectangular prism with a height of 32 cm, a length of 18 cm, and a width of 6 cm?

A. 6912 cm^3 B. 1536 cm^3
C. 1728 cm^3 D. 3456 cm^3

14. What is the volume of the cylinder? Round your answer to the nearest tenth. Leave your answer in terms of π.

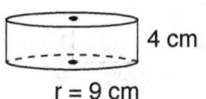

A. 1296.0π cm^3 B. 432.0π cm^3
C. 648.0π cm^3 D. 324.0π cm^3

15. What is the volume of the oblique cylinder? Round your answer to the nearest tenth. Leave your answer in terms of π.

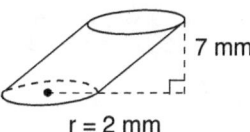

A. 28.0π mm^3 B. 29.1π mm^3
C. 38.8π mm^3 D. 39.8π mm^3

CHAPTER 6
Probability, Statistics, and Data Analysis

In this chapter, you will learn how to find the range, mean, median, and mode of data sets. Also, you will learn about statistics, data analysis, and probability.

1. Mean, Median, and Mode

6–1. Mean, Median, and Mode

> Mean: The average number in a data set that is found by adding up the scores in the data and dividing it by the total number of scores.
> Median: The number located in the middle of a data set put in numerical order.
> Mode: The number that occurs most often in the data set.
> Range: The range is the difference between the largest and the smallest values in a data set.

6–2. How does Δ affect the mean and median?

$$73, 67, 71, 71, \text{ and } \Delta$$

a) Mean
b) Median

SOLUTION

a) The mean (average) of the known numbers is 70.5. So if Δ is less than 70.5, the mean will decrease. However, if Δ is greater than 70.5, the mean will increase.

b) The median will not changed if Δ is less than 70.5 nor if Δ is greater than 70.5.

*Solving Problems

Exercises 1   Solve each problem using the given information.

1) Joan takes a vocabulary quiz every Friday at school. She recorded her scores as 12, 14, 13, 13, 15, 13, 15, and 14. What are the mode and median for those scores?

2) Chai is making a survey of her friends about how many books they reads every day for a week. She found that 3.5 books is the mean, 2 books is the mode, and 2.5 books is the median. Explain why the mean is higher than the median.

Exercises 2   The table below shows a set of data that Shawn made about the recyclables that are collected weekly.

Weekly Collected Recyclables

Item \ Week	1	2	3	4	5
Plastic Bottles	25	27	34	24	25
Cans	39	25	44	21	29

1) If Shawn compared the number of plastic bottles and cans, what would be the best kind of graph in order to understand the data set?

2) Construct a stem-and-leaf plot based on the data set.

3) What are the mean, mode, median, and range of the data?

4) If Shawn collected recyclables for one more week, the mean of the data set is increased. How is the median affected?

5) If Shawn collected recyclables for one more week, the median of the data set is increased. How is the mean affected?

Exercises 3   Use the information for Exercises **1-3**. Paul recorded the hours he spent working on his math and English.

Weekly Working Homework

Week / Subject	1	2	3	4
English	175	205	185	190
Math	230	250	230	235

1) Construct a stem-and-leaf plot based on the data set.

2) What are the mean, mode, median, and range of the data set?

3) Paul added an additional week, but the median was decreased in the data set. How was the mean affected?

SELF-TEST

Use the information for Exercises **1-3**. Lily recorded the temperature each day before school started.

Weekly Temperature (°F)

Monday	76
Tuesday	74
Wednesday	85
Thursday	77
Friday	76

1. What is the mean of the temperatures?

   A. 75   **B.** 77.6
   C. 76.5   **D.** 76

2. What is the median of the temperatures?

    A. 74   **B.** 75
    C. 76   **D.** 77

3. On Saturday, if the temperature suddenly increased $10°$ F more than the temperature on Wednesday and she added it to the data set, which of the following statements is correct?

    A. The mode increases.   **B.** The median increases.
    C. The median does not change.   **D.** The mean does not change.

4. The mean in the data set on all 5 is 24.5. What is the unknown number?
          19.5, 24, ____, 26.5, 26

    A. 24   **B.** 24.5
    C. 25   **D.** 26.5

5. The median and mean in the data set are both 131. What is the unknown number?
          142, 112, 150, ____, 124, 138

    A. 131   **B.** 125
    C. 110   **D.** 120

6. If the mean in the data set is increased, what is the unknown number?
          12, ____, 15, 9

    A. 8   **B.** 10
    C. 12   **D.** 14

7. If the mean in the data set is 3, what is the unknown number?
          1, 5, ____, 5, 3

    A. 1   **B.** 2
    C. 3   **D.** 5

2. Probability

6-3. Know the concepts of probability with one event.

> * Probability with one event
>
> - Probability is the chance that something will happen.
> - An outcome is the possible result of an experiment.
> - An event is a set of outcomes of an experiment.
>
> $$\text{Probability } (P) \text{ of an event} = \frac{\text{number of successful outcomes}}{\text{total number of possible outcomes}}$$

6-4. A box has 4 red marbles, 2 blue marbles, 6 yellow marbles, and 3 violet marbles. If you take out a random marble and put it back in after checking it, find the probabilities for the following colors:

    a) red   b) yellow   c) blue or violet   d) orange

SOLUTION

The box contains 15 marbles in total.
a) There are 4 red marbles in the box.

$$P \text{ (red)} = \frac{4}{15} \quad \longleftarrow \text{ number of successful outcomes (4 red)}$$
$$\phantom{P \text{ (red)} = \frac{4}{15}} \quad \longleftarrow \text{ total number of possible outcomes (4 red + 2 blue + 6 yellow +3 violet)}$$

b) There are 6 yellow marbles in the box.

$$P \text{ (yellow)} = \frac{6}{15} \quad \longleftarrow \text{ number of successful outcomes (6 y)}$$
$$\phantom{P \text{ (yellow)} = \frac{6}{15}} \quad \longleftarrow \text{ total number of possible outcomes (4 r + 2 b + 6 y +3 v)}$$

$$P \text{ (yellow)} = \frac{2}{5} \qquad \text{Divide 6 and 15 with the GCF of 3.}$$

c) There are 2 blue marbles and 3 violet marbles in the box.

$$P \text{ (blue or violet)} = \frac{5}{15} \quad \longleftarrow \text{ number of successful outcomes (2 b and 3 v)}$$
$$\phantom{P \text{ (blue or violet)} = \frac{5}{15}} \quad \longleftarrow \text{ total number of possible outcomes (4 r + 2 b + 6 y +3 v)}$$

$$P \text{ (blue or violet)} = \frac{1}{3} \qquad \text{Divide 5 and 15 with the GCF of 5.}$$

d) There are no orange marbles in the box.
$$P \text{ (orange)} = 0$$

6-5. Probability for two events

> In order to find the probability for two events, there are three possible methods, which are systematic listing, using a table, or using a diagram.

6-6. There are 12 chocolate ice cream cones, 9 vanilla ice cream cones and 3 strawberry ice cream cones. Choosing an ice cream cone at random, what is the percentage of the probability of choosing a vanilla cone?

SOLUTION

There are 9 vanilla ice cream cones. So, first, set up the ratio of number of successful outcomes with total number of possible outcomes and use the equation below to convert the fraction into a percentage.

*Let x represent the unknown number.

$$\boxed{\textbf{Fraction} \times \textbf{100\%} = x}$$

* Remember how to turn a fraction into a percentage

$\frac{9}{24} \times 100\% = x$ Substitute the equation with $\frac{9}{24}$.

$\frac{3}{8} \times 100\% = x$ Divide 9 and 24 with the GCF 3.

$3 \times 12.5\% = x$ Divide 3 by 8.
$37.5\% = x$ Multiply 3 by 12.5.

So there is a 37.5% probability of choosing a vanilla ice cream cone.

6-7. A ice cream store sold chocolate ice cream, vanilla ice cream, and strawberry ice cream. A customer can also choose small, medium, or large.

a) What is the probability that a customer will choose a small chocolate ice cream?
b) What percentage is the probability that a customer can choose any medium sized ice cream?
c) How many possible outcomes of sizes and flavors are there?

SOLUTION

Solution
a) There are 9 possible outcomes with 3 flavors and 3 different sizes. So the probability of a medium size is the product of $\frac{1}{3}$ of flavors and $\frac{1}{3}$ of sizes.

P (medium) $= \frac{1}{3} \times \frac{1}{3} = \frac{1}{9}$, so you can convert the fraction as a percentage. Let x

represents the unknown number

$$\frac{1}{9} \times 100\% = x \qquad \text{Substitute } x \text{ with } \frac{1}{9}.$$

$$11\% \approx x \qquad \text{Divide 100 by 9.}$$

So the probability is about 11%.

b) There are 3 choices of a medium size chocolate, vanilla, or strawberry ice cream cone out of 9 possible choices. So the probability is $\frac{3}{9}$, or $\frac{1}{3}$, or about 33%.

c) The possible outcomes are shown below.

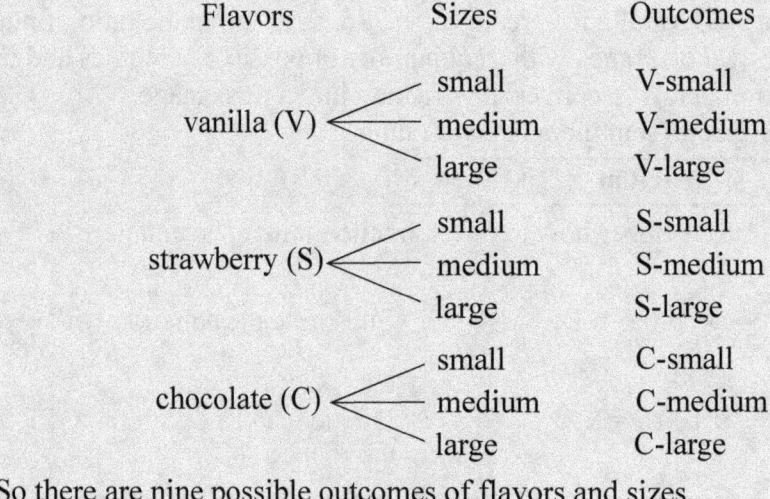

So there are nine possible outcomes of flavors and sizes.

6–8. A customer can choose the items listed on the menu below.

List 1	List 2
Hamburger	Ketchup
Hot dog	Mustard
Sandwich	

a) What are the possible combinations that a customer can choose?
b) How many possible combinations that a customer can choose?

SOLUTION

a) Lists 1 and 2 have Hamburgers (Hb), Hot dogs (Hd), Sandwiches (S), Ketchup (K), and Mustard (M). So the possible combinations are (Hb and K), (Hb and M), (Hd and K), (Hd and M), (S and K), and (S and M).
b) The total number of combinations that a customer can choose is 6.

* Solving Problems

Exercises 4   Use the following information for Exercises **1-4**. An ice cream store sells chocolate ice cream, vanilla ice cream, and strawberry ice cream. A customer can either choose a cup or a cone to put the ice cream in. Each cup and cone comes in small, medium, or large.

1)   What is the total number of different items that a customer can choose?

2)   What is the probability that a customer will choose the large vanilla ice cream cup?

3)   What is the percentage of the probability that a customer will choose a small size ice cream?

4)   What are the possible outcomes?

Exercises 5   Use the following information for Exercises **1-2**. A customer can choose the items listed on the menu below.

List 1	List 2
Chicken Hamburger	Cheese
Beef Hamburger	No cheese
Chicken Sandwich	Onions
Beef Sandwich	No Onions
	French Fries

1)   What is the total number of combinations that customers can choose if they select one item from each list?

2)   What is the total number of combinations that customers can choose, given that they choose one item from List 1 and 3 items from List 2?

Exercises 6   Use the following information for Exercises **1-3**. A card dealer has 16 cards laid out. There are four sets of hearts, clover, diamonds, and spades. Each card is a king, queen, jack, or ace.

1)   What is the probability of choosing an ace of diamonds?

2)   What is the percentage of the probability of choosing a queen?

3)   What is the probability of choosing a king of clover and any jack?

Exercises 7   Use the following information for Exercises **1-3**. There are 3 dices on a table. One die has 1-6, the second die has 7-12, and the third die has A-F.

1)   What is the probability of rolling a 5 and an A?

2)   What is the percentage of the probability of rolling a multiple of 4 and an A or C?

3)   Find the possible outcomes of the rolled dices.

Exercises 8   There is a website online that sells shirts. Each shirt comes in small, medium, and large. In addition, each t-shirt comes in red, blue, green, yellow, and orange. A customer can either choose a short-sleeved t-shirt or a long-sleeved t-shirt.

1)   What is the probability of a customer choosing a medium orange long-sleeved shirt?

2)   What is the percentage of a customer choosing any sized orange short-sleeved shirt?

3)   How many possible outcomes of t-shirts are there?

SELF-TEST

Use the information for Exercises **1-2**. A box has 7 red candies, 3 blue candies, and 9 yellow candies.

1. If you take out a random candy, what is the probability of choosing a blue candy?

    A. $\dfrac{1}{3}$ **B.** $\dfrac{3}{19}$

    C. $\dfrac{3}{10}$ **D.** $\dfrac{3}{189}$

2. If you take out a random candy, what is the percentage of the probability of choosing a yellow candy?

    A. 15.8%   **B.** 36.8%
    C. 47.4%   **D.** 5.3%

Use the following information for Exercises **3-5**. When making a sandwich, a person can choose between white and whole wheat bread, turkey and ham, and Swiss and cheddar.

3. What is the probability of a person choosing a turkey sandwich with white bread?

    A. $\dfrac{1}{3}$ **B.** $\dfrac{1}{4}$

    C. $\dfrac{1}{2}$ **D.** $\dfrac{2}{9}$

4. What is the percentage of the probability of choosing a sandwich with cheddar in it?

    A. 25%   **B.** 33%
    C. 58%   **D.** 66%

5. How many possible combinations of sandwiches are there?

    A. 3   **B.** 6
    C. 9   **D.** 12

Use the following information for Exercises **6-8**. There is a sale on phone cases. Each phone case comes in white, black, and red and has either stripes or polka dots.

6. What is the probability of choosing a red striped phone case?

    A. $\dfrac{1}{2}$ **B.** $\dfrac{1}{3}$

    C. $\dfrac{1}{4}$ **D.** $\dfrac{1}{6}$

7. What is the percentage of the probability of choosing a case with polka dots?

      A. 10%   **B.** 25%

      C. 66%   **D.** 33%

8. Find the possible combinations of phone cases.

      A. 4   **B.** 5

      C. 6   **D.** 7

Use the following information for Exercises **9-11**. For Adam's birthday, his family wants to buy him a cake. They can choose a chocolate or vanilla cake, with 3 different frostings of chocolate, vanilla or strawberry and can choose to have a circular or rectangular cake.

9. What is the probability of choosing a rectangular chocolate cake with strawberry frosting?

      A. $\dfrac{1}{3}$ **B.** $\dfrac{1}{6}$

      C. $\dfrac{1}{4}$ **D.** $\dfrac{1}{12}$

10. What is the percentage of the probability of getting a circular cake with vanilla frosting?

      A. 15%   **B.** 25%

      C. 33%   **D.** 50%

11. Find the possible combinations of cakes.

      A. 6   **B.** 9

      C. 10   **D.** 12

Use the information for Exercises **12-14**. The diagram is a spinner showing odd numbers from 1-15.

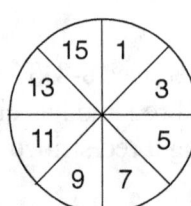

12. When you spin the spinner two times, what are the probabilities of getting 1 and 9?

A. $\frac{1}{16}$ B. $\frac{1}{64}$

C. $\frac{1}{26}$ D. $\frac{1}{32}$

13. What is the probability of spinning a number divisible by 3?

A. $\frac{1}{9}$ B. $\frac{1}{8}$

C. $\frac{1}{16}$ D. $\frac{3}{8}$

14. What is the percentage of the probability of spinning a number divisible by 5?

A. 12.5%   B. 25%
C. 50%   D. 62.5%

Use the information for Exercises **15-17**. The diagram is a spinner showing various letters.

15. What is the probability of spinning a vowel?

A. $\frac{1}{2}$ B. $\frac{1}{3}$

C. $\frac{1}{4}$ D. $\frac{1}{8}$

16. What is the probability of spinning A or O?

A. $\frac{1}{2}$ B. $\frac{1}{3}$

C. $\frac{1}{4}$ D. $\frac{1}{8}$

17. What is the percentage of the probability of spinning the letters MIKA?

A. 12.5%   B. 25%
C. 50%   D. 62.5%

ANSWERS

CHAPTER 1

Quick Exercises 1
1) 6   2) 2

Quick Exercises 2
1) 7   2) −10

Quick Exercises 3
1) −8   2) −15

Quick Exercises 4
1) −3   2) 5

Quick Exercises 5
1) 2/5   2) 1/3

Quick Exercises 6
1) $30 = 30$ 2) $35 \neq 27$

Quick Exercises 7
1) 1:2.5   2) 1:35

Quick Exercises 8
1) $3/2 = x$ 2) $x = 3$ 3) $x = 9$ 4) $x = 1$

Exercises 1
1) 0   2) −2   3) 4   4) −4   5) 18   6) −$23.50
7) −42   8) −63   9) −39   10) 14   11) −15   12) −14
13) 5   14) −15

Exercises 2
1) 17   2) 38   3) 40   4) −16   5) −18   6) −71
7) $7.00   8) −7   9) 9   10) 3   11) −23   12) −12

Exercises 3
1) $67.00   2) −0.43   3) 8.3   4) −11   5) −$2.00   6) −3
7) 31   8) −23   9) 2   10) 0   11) 9   12) −15

Exercises 4
1) 96   2) −0.08   3) −104   4) 70   5) −0.5   6) −90
7) $-1\frac{3}{5}$ 8) 56

Exercises 5
1) −4   2) −18   3) −21.25   4) −37.5   5) −$32.00   6) −$57.75
7) 9   8) 20   9) −60   10) −9   11) −168   12) 2
13) 24   14) −0.5

Exercises 6
1) −15   2) 0.25   3) −96   4) 13   5) −2.72   6) 13.6
7) 24   8) 1/3   9) −20   10) 9

Exercises 7
1) $231 \neq 224$ 2) $42 = 42$

Exercises 8
1) 242 = 48x   2) 18x = 30   3) 15x = 16   4) 4x = 5
    121/24 = x   x = 3/5   x = 16/15   x = 5/4

Exercises 9
1) 9:4   2) 3.5:0.6   3) 6:5   4) 2:3   5) 4:0.25   6) 0.029:0.3
7) 5:6   8) 0.41:0.6   9) 9:4   10) 4:3   11) 1:1   12) 0.02:1.5

Exercises 10
1) 66/7   2) 3/2   3) 16   4) 20   5) 40   6) 1/3
7) 1   8) 2   9) 26/3   10) 3   11) 3   12) 10

Exercises 11
1) 85 min   2) $52.00   3) $9.20   4) $6.00   5) $9.60   6) $3.15

Exercises 12
1) x = 2 cm   2) x = 36 units   3) 3

Self-Test, Page 9
1) D   2) C   3) D   4) A   5) C   6) A
7) C   8) A   9) D   10) A   11) B

Self-Test, Page 13
1) B   2) A   3) B   4) D   5) B   6) C
7) D   8) C   9) D   10) C

CHAPTER 2

Quick Exercises 1   Quick Exercises 2
1) -6   2) -30   1) 6   2) 20

Quick Exercises 3   Quick Exercises 4
1) 4   2) –8   1) $1\frac{1}{2}$ 2) 12

Quick Exercises 5   Quick Exercises 6
1) 8   2) 9   1) 59.9   2) 38.4

Quick Exercises 7   Quick Exercises 8
   $y = 3x$ $y = x - 1$

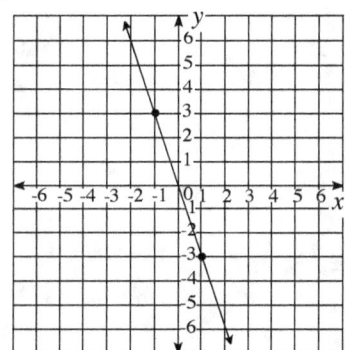   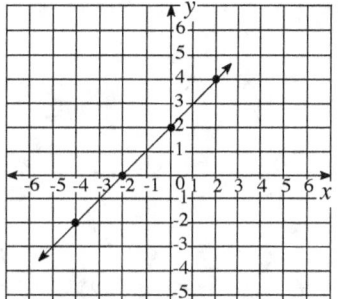

Quick Exercises 9
1) slope: 7, y-inter: –2   **2)** slope: 4, y-inter: 1

Quick Exercises 10
1) slope: –7, y-inter: –2   **2)** slope: 2, y-inter: $-\frac{1}{2}$

Quick Exercises 11   Quick Exercises 12
1) slope: –2, y-inter: 0   **2)** slope: $\frac{1}{5}$, y-inter: –1   **1)** slope: $\frac{1}{2}$, y-inter: –2

Exercises 1
1) 4   **2)** –11   **3)** 10   **4)** 25

Exercises 2
1) –13   **2)** –1   **3)** 231   **4)** 26.22   **5)** 48   **6)** 50
7) –1   **8)** 126   **9)** 12   **10)** 5

Exercises 3
1) –15   **2)** 1   **3)** –5   **4)** 9   **5)** $-1\frac{2}{3}$ **6)** $-\frac{1}{3}$
7) –2   **8)** 1

Exercises 4
1) –16   **2)** 2   **3)** $-4\frac{1}{2}$ **4)** $5\frac{3}{4}$ **5)** 28   **6)** 32
7) 60   **8)** –12   **9)** $-1\frac{1}{4}$ **10)** –2

Exercises 5
1) 1   **2)** $\left(\frac{2}{3}\right)$ **3)** $\left(\frac{1}{2}\right)$ **4)** 4

Exercises 6
1) 7   **2)** –12   **3)** 4   **4)** 48   **5)** 2   **6)** 50
7) 64   **8)** 80   **9)** –14   **10)** –30   **11)** –60   **12)** –6

Exercises 7
1) $3\frac{1}{4}$ **2)** $\left(\frac{3}{5}\right)$ **3)** $\left(\frac{1}{3}\right)$ **4)** 6   **5)** 3   **6)** –2

Exercises 8
1) –1   **2)** $-1\frac{2}{3}$ **3)** –3   **4)** 3   **5)** –1   **6)** –6
7) –2   **8)** –8

Exercises 9
1) 1   **2)** $1\frac{1}{2}$ **3)** $-\frac{5}{9}$ **4)** $-\frac{1}{3}$ **5)** $1\frac{1}{4}$ **6)** 3
7) –3   **8)** 4

Exercises 10
1) $\frac{430}{40}(x) = 1290$ **2)** 120

Exercises 11
1) $5 + 12 + 3 + 15 + x = 8{:}30$ **2)** 7:55am   **3)** 8:05am   **4)** 7:35am

Exercises 12
1) −24   2) −6   3) −0.1   4) −10   5) 3.2   6) 1.2
7) −2.1   8) 8   9) 8   10) −1   11) −2   12) −4

Exercises 13
1) $-\dfrac{1}{2}$ 2) $\left(\dfrac{1}{2}\right)$ 3) 0.8   4) −7   5) 6   6) 24

Exercises 14
1) −6   2) −6   3) −0.4   4) −11   5) 384   6) 36
7) −70   8) 21   9) $\left(\dfrac{1}{2}\right)$ 10) −2   11) 63   12) −4

Exercises 15
1) 8   2) 1.4   3) 32   4) 5   5) −4   6) 10

Exercises 16
1) 6   2) 17   3) 260.4   4) 0.1   5) −5   6) −9
7) 6   8) 0.3   9) 1.5   10) −0.6   11) 2   12) −0.5

Exercises 17
1) 5   2) −1.9   3) 12.8   4) −24   5) 6   6) −14

Exercises 18
1) −2   2) −1.5   3) −60   4) −26   5) −3   6) −48
7) 50   8) 2   9) $4\dfrac{1}{2}$ 10) $-1\dfrac{2}{3}$ 11) $-\dfrac{2}{3}$ 12) −4

Exercises 19
1) 3   2) 3   3) $\left(\dfrac{1}{2}\right)$ 4) $\left(\dfrac{2}{3}\right)$ 5) $3\dfrac{1}{3}$ 6) $\left(\dfrac{2}{3}\right)$
7) 2   8) 1   9) 2   10) $6\dfrac{1}{3}$

Exercises 20
1) −0.14   2) −0.5   3) −2   4) −71.5   5) 8   6) 1

Exercises 21
1) 20%   2) $18.84

Exercises 22
1) $103 = 8x - 1$ 2) 13

Exercises 23
1) $410 = 35x - 2(5)$ 2) 12

Exercises 24
1) $0.25(x - 7) + 0.1y$ 2) $x = 1.25, y = 1.2$ 3) 2.45

Exercises 25
1) $8x + (x + 25) = 340$ 2) $35.00

Exercises 26
1) 4   2) 15   3) $4\dfrac{1}{2}$ 4) −2   5) 18   6) 1

Exercises 27

1) 20 **2)** 6 **3)** 4 **4)** 19 **5)** $-\dfrac{1}{2}$ **6)** 3

Exercises 28

1) $0.75x + 0.55y$ **2)** $0.75x + 0.55(5) = 12.50$ **3)** $0.75x + 0.55x(4) = 12.50$

Exercises 29

1) 1.1514 **2)** 4 **3)** 1.1 **4)** 206 **5)** 384 **6)** 115.25
7) 1.68 **8)** 180 **9)** 1.641 **10)** 42

Exercises 30(A)

1) 7.524 **2)** 204 **3)** 846 **4)** 81 **5)** 248 **6)** 76.322
7) 4320 **8)** 192 **9)** 108 **10)** 3.672

Exercises 30 (B)

1) 7.524 **2)** 204 **3)** 846 **4)** 81 **5)** 248 **6)** 76.322
7) 4320 **8)** 192 **9)** 108 **10)** 3.672

Exercises 31

1) $y = 3x$

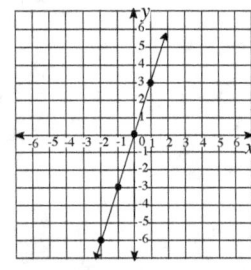

2) $y = -x$

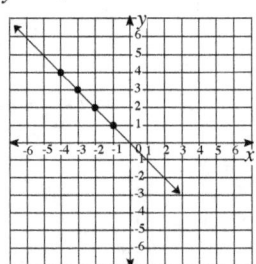

3) $y = -3x$

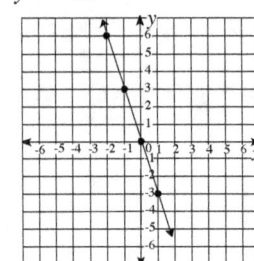

4) $y = 2x$

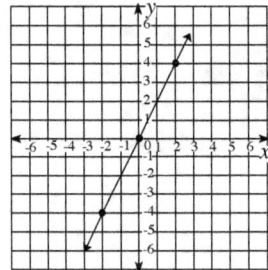

Exercises 32

1) $y = x + 1$

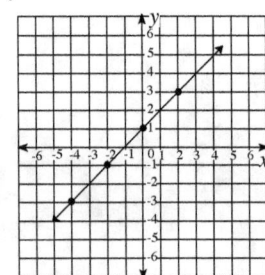

2) $y = -x + 1$

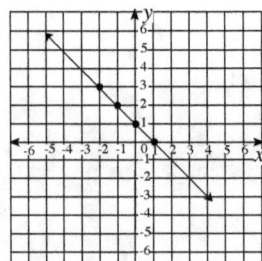

3) $y = 2x + 1$

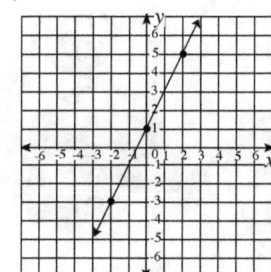

4) $y = 3x + 2$

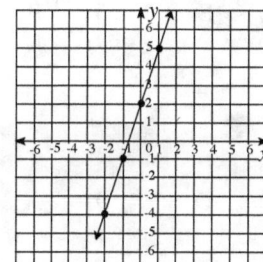

Exercises 33

1)   slope: 3, y-inter: -1

2)   slope: 3, y-inter: $-\frac{1}{2}$

3)   slope: 1, y-inter: 4

4)   slope: 4, y-inter: 3

5)   slope: -7, y-inter: -3

6)   slope: -5, y-inter: 0

7)   slope: 2, y-inter: 0

8)   slope: 0, y-inter: -1

9)   slope: $\frac{1}{2}$, y-inter: -3

10)   slope: -5, y-inter: $\frac{1}{2}$

Exercises 34

1) $y = 2x + 1$

2) $y = -2x - 1$

3) $y = -x + 5$

4) $y = 5x - 2$

5) $y = 3x$

6) $y = -2x + 3$

Exercises 35

1)   slope: $-\frac{2}{3}$, y-inter: -1

2)   slope: -3, y-inter: -3

3)   slope: 2, y-inter: -2

4)   slope: -1, y-inter: 4

5)   slope: -1, y-inter: 5

6)   slope: $\frac{2}{5}$, y-inter: 2

Exercises 36

1) $y = \frac{1}{2}x - 3$

2) $y = -\frac{2}{3}x - 4$

3) $y = -6x + 6$

4) $y = 2x + 4$

Exercises 37

1) $y = 5x + 5$

2) $y = x + 1$

3) $y = x - 2$

4) $y = 2\frac{1}{2}x + 1$

Self-Test, Page 23

1)	D	**2)**	A	**3)**	B	**4)**	C	**5)**	A	**6)**	A
7)	D	**8)**	C	**9)**	B	**10)**	B	**11)**	C	**12)**	A
13)	B	**14)**	C	**15)**	C	**16)**	B	**17)**	D	**18)**	C
19)	D	**20)**	D	**21)**	B	**22)**	A	**23)**	B	**24)**	C
25)	D										

Self-Test, Page 35

1)	D	**2)**	C	**3)**	C	**4)**	A	**5)**	B	**6)**	B
7)	B	**8)**	A	**9)**	A	**10)**	A	**11)**	A	**12)**	A
13)	D	**14)**	D	**15)**	C	**16)**	A	**17)**	C	**18)**	B
19)	C	**20)**	D	**21)**	A	**22)**	B	**23)**	C	**24)**	B
25)	C	**26)**	B	**27)**	D	**28)**	B	**29)**	B	**30)**	A

Self-Test, Page 47

1)	B	**2)**	D	**3)**	C	**4)**	B	**5)**	C	**6)**	A
7)	B	**8)**	D	**9)**	A	**10)**	C	**11)**	D		

1) D 2) B 3) B 4) B 5) D 6) C
7) C 8) D 9) B 10) A 11) C 12) A
13) C

1) B 2) C 3) A 4) C 5) D 6) B
7) C 8) D 9) A 10) A 11) B 12) C

CHAPTER 3

Quick Exercises 1
1) −2.00 2) 1.00

Quick Exercises 2
1) −0.06 < 0.06 2) −5.06 > -5.60

Quick Exercises 3
1) −0.7 2) −7.69

Quick Exercises 4
1) 5.93 2) −0.51

Quick Exercises 5
1) −3.75 2) 0.888 3) 1.52 4) −0.21

Quick Exercises 6
1) −32.6 2) 4.9 3) -0.74 4) 37.2

Quick Exercises 7
1) 1.386 2) 8.012 3) −0.115 4) −0.072

Quick Exercises 8
1) −2.153 2) 3.104

Quick Exercises 9
1) 5 2) 3

Quick Exercises 10
1) $3\frac{1}{2}$ 2) $3\frac{1}{8}$

Quick Exercises 11
1) $4\frac{4}{9}$ 2) $3\frac{7}{10}$

Quick Exercises 12
1) 4 2) 10 3) $(\frac{1}{2})$ 4) $-\frac{2}{3}$

Quick Exercises 13
1) $(\frac{1}{2})$ 2) $1(\frac{7}{25})$

Quick Exercises 14
1) $1\frac{1}{2}$ 2) $1\frac{3}{4}$

Quick Exercises 15
1) $(\frac{1}{2})$ 2) 2

Exercises 1
1) 3.00 2) 53.00 3) 558 4) 16 5) −47 6) −15
7) 34 8) −3 9) 47 10) −9

Exercises 2
1) −0.01 < 0 2) 0.92 > −1.50 3) 0.42 > −0.36 4) −2.03 > −3.52
5) −0.74 > −0.84 6) −1.06 < 0.06 7) −0.29 < −0.08 8) −5.06 < −5.04
9) −1.06 < 0.04 10) 0.1 > 0.04 11) 4.075 < 4.10 12) −0.08 < 0.01

Exercises **3**
1) $-1.02 < -0.14 < -0.04 < 0.04 < 0.14$
2) $-0.3 < -0.25 < -\frac{1}{5} < 0 < \frac{2}{7}$
3) $-\frac{6}{9} < -0.625 < -\frac{4}{7} < -\frac{4}{8}$
4) $-1.33 < -1.3 < -1\frac{2}{10} < -1.03 < -0.99$
5) $-\frac{1}{6} < -\frac{1}{8} < -\frac{1}{9} < -0.09$
6) $-2\frac{1}{4} < -1\frac{1}{2} < -1.09 < 1\frac{1}{3}$
7) $0.014 < 0.04 < 0.14 < 1.02$
8) $-\frac{1}{4} < -\frac{8}{100} < 0 < \frac{9}{10}$

Exercises **4**
1) -1.4 **2)** -1.0 **3)** -0.7 **4)** -0.6

Exercises **5**
1) -9.2 **2)** -8.9 **3)** -7.7 **4)** -7.4

Exercises **6**
1) -4.2 **2)** -1.9 **3)** -0.6 **4)** 0.6

Exercises **7**
1) -0.45 **2)** -0.35 **3)** -0.2 **4)** -0.15

Exercises **8**
1) 4.48 **2)** -0.84 **3)** -7.96 **4)** 1.2 **5)** -2.52 **6)** 37.35
7) -62.65 **8)** -15.68

Exercises **9**
1) -0.747 **2)** -0.197 **3)** -1.198 **4)** 0.009 **5)** -0.997 **6)** -0.034
7) -515 **8)** 3.226 **9)** 26.825 **10)** 57.97 **11)** 22.11 **12)** 32.736
13) 36.022

Exercises **10**
1) -0.59 **2)** -4.19 **3)** 2.89 **4)** 1.48 **5)** 2.04 **6)** 0.29

Exercises **11**
1) 91.39 **2)** -0.51 **3)** 2.61 **4)** -5.436 **5)** -0.242 **6)** 3.862
7) 74.603 **8)** 28.25 **9)** 28.999 **10)** 2.863 **11)** 46.905

Exercises **12**
1) -1.4 **2)** -2.73 **3)** 2 **4)** 1.8 **5)** 1.2 **6)** 2.0

Exercises **13**
1) 0.446 **2)** 0.103 **3)** 15.032 **4)** -1.238 **5)** -2.23 **6)** 7.492
7) -2.016 **8)** -10.04

Exercises **14**
1) 0.14 **2)** -1.22 **3)** 5.94 **4)** 4.95 **5)** 3.72 **6)** 1.09
7) 11.38 **8)** 2.43 **9)** 4.02 **10)** -2.97 **11)** -5.24 **12)** 12.55

Exercises **15**
1) $15x = 187.95$ **2)** 25.24 **3)** 13.61

Exercises **16**
1) -20.00 **2)** -0.94 **3)** 0.78 **4)** -3.6

Exercises 17
1) 106.7292   2) −25.2   3) −22.86   4) 19   5) 24.6   6) 14.4
7) 0.225   8) 31.248   9) −0.533   10) −0.096   11) 1067.64   12) 288.5596
13) 488.2612

Exercises 18
1) −5.7   2) −0.82   3) 1.25   4) 3.9   5) −31.08   6) −0.34

Exercises 19
1) 11.04   2) 0.0882   3) −3   4) −1.856   5) −326.167   6) −18.27
7) −0.9895   8) 26.0104

Exercises 20
1) 0.07   2) −0.41   3) 1.98   4) 0.99   5) −5.44   6) 0.83
7) 1.26   8) 0.6   9) 10.94   10) 16.65

Exercises 21
1) 31   2) 63.6   3) −4.5   4) 42.6   5) −7.56   6) −1.3

Exercises 22
1) 18.6   2) 6.5   3) 5.6   4) 0.17   5) −1.1   6) −4.5
7) −9.2   8) −1.3   9) 7.5   10) −0.24   11) 0.432   12) 5.3

Exercises 23
1) −0.24   2) −6.1   3) −24.5   4) −6.7   5) 4.5   6) 6.5
7) −0.25   8) −0.0525

Exercises 24
1) 2.5   2) −2.6   3) 5   4) 16   5) 1.0   6) 0.18

Exercises 25
1) 1.2   2) 0.9   3) 0.168   4) −18.9   5) 3.1   6) 0.03
7) −3   8) 0.4   9) 2.24   10) −2.46   11) −12.5   12) 8.602

Exercises 26
1) 11.87 inches   2) 6.3 inches   3) $18.675 = 0.45x, x = 41.5$

Exercises 27
1) 2.49   2) 2h27 m   3) 2.81lb

Exercises 28
1) $\frac{2}{5} > \frac{3}{8}$ 2) $\frac{3}{4} < \frac{5}{7}$ 3) $\frac{3}{6} < \frac{4}{5}$ 4) $\frac{4}{3} < \frac{7}{10}$
5) $\frac{8}{9} > 1.25$ 6) $0.95 < \frac{9}{10}$ 7) $2.15 < 3\frac{1}{4}$ 8) $\frac{1}{3} < \frac{1}{4}$
9) $\frac{5}{6} > \frac{6}{7}$ 10) $\frac{1}{4} < 0.05$ 11) $-1\frac{1}{2} < -1\frac{1}{3}$ 12) $0.12 < -1\frac{3}{4}$

Exercises 29
1) $\frac{3}{4} > \frac{2}{3} > \frac{1}{2} > \frac{1}{8}$ 2) $\frac{2}{3} > \frac{5}{8} > \frac{4}{7} > -0.6$
3) $\frac{3}{4} > \frac{3}{5} > \frac{1}{2} > \frac{1}{3}$ 4) $\frac{5}{8} > \frac{2}{5} > 0.375 > 0.309$
5) $\frac{2}{5} > \frac{1}{3} > 0.25 > -0.3$ 6) $1\frac{1}{16} > 1.25 > -1\frac{1}{9} > -1.25$

7) $-\frac{1}{6} > -\frac{1}{5} > -\frac{1}{4} > -\frac{1}{3}$　　　　**8)** $-1\frac{1}{3} > -2\frac{1}{4} > -3\frac{1}{5} > -4\frac{1}{6}$

Exercises 30
1) $-\frac{2}{5}$　　　**2)** $-\frac{1}{5}$　　　**3)** $(-\frac{1}{10})$　　　**4)** $-\frac{2}{5}$

Exercises 31
1) $-3\frac{1}{3}$　　　**2)** $-\frac{1}{10}$　　　**3)** $-\frac{4}{5}$　　　**4)** $1\frac{3}{10}$

Exercises 32
1) $-3\frac{1}{10}$　　　**2)** $-3\frac{3}{10}$　　　**3)** $-3\frac{6}{10}$ or $-3\frac{3}{5}$　　　**4)** $-3\frac{4}{5}$

Exercises 33
1) -1　　　**2)** -2　　　**3)** $-2\frac{1}{3}$　　　**4)** -3

Exercises 34
1) 3　　**2)** 5　　**3)** -1　　**4)** 2　　**5)** 1　　**6)** -1

Exercises 35
1) -1　　**2)** -1　　**3)** $(\frac{4}{7})$　　**4)** -2　　**5)** $2\frac{3}{4}$　　**6)** $-\frac{1}{6}$
7) $(\frac{39}{56})$　　**8)** $(\frac{1}{3})$

Exercises 36
1) $(\frac{1}{2})$　　**2)** $-\frac{1}{6}$　　**3)** $(-\frac{1}{10})$　　**4)** $-\frac{1}{2}$　　**5)** 5　　**6)** -1
7) -10　　**8)** -2

Exercises 37
1) $-\frac{7}{10}$　　**2)** $-\frac{7}{8}$　　**3)** $(\frac{5}{8})$　　**4)** $-2\frac{1}{12}$　　**5)** $-\frac{5}{6}$　　**6)** $-\frac{2}{9}$
7) $(\frac{1}{6})$　　**8)** $-1\frac{9}{10}$

Exercises 38
1) $-\frac{1}{6}$　　**2)** $(\frac{1}{3})$　　**3)** $1\frac{7}{12}$　　**4)** $2\frac{1}{21}$　　**5)** $-\frac{1}{6}$　　**6)** $-\frac{1}{2}$

Exercises 39
1) $(\frac{2}{3})$　　**2)** $-2\frac{1}{20}$　　**3)** $(\frac{1}{15})$　　**4)** $(\frac{1}{9})$　　**5)** $-\frac{14}{15}$　　**6)** $3\frac{4}{7}$

Exercises 40
1) $-\frac{1}{9}$　　**2)** $1\frac{1}{2}$　　**3)** $-1\frac{1}{3}$　　**4)** $-\frac{2}{3}$　　**5)** $-1\frac{1}{3}$　　**6)** $(\frac{4}{5})$

Exercises 41
1) $1\frac{1}{6}$　　**2)** $3\frac{1}{2}$　　**3)** $1\frac{1}{2}$　　**4)** -16

Exercises 42
1) $(\frac{5}{6})$　　**2)** $1\frac{1}{4}$　　**3)** $(\frac{1}{8})$　　**4)** $-1\frac{3}{8}$　　**5)** $-\frac{5}{6}$　　**6)** $-2\frac{3}{8}$

Exercises 43
1) $1\frac{1}{4}$　　**2)** $-\frac{6}{7}$　　**3)** $-\frac{1}{3}$　　**4)** $-1\frac{3}{8}$　　**5)** $(\frac{1}{2})$　　**6)** $(\frac{1}{2})$

Exercises 44
1)   1   2)   4   3)   −4   4)   −2

Exercises 45
1) $(\frac{1}{16})$ 2) $(\frac{6}{25})$ 3) $(\frac{1}{48})$ 4) $(\frac{2}{3})$ 5) $-1\frac{3}{5}$ 6) $-\frac{1}{3}$

Exercises 46
1) $-\frac{1}{2}$ 2) −2   3) $-\frac{1}{5}$ 4) $-2\frac{1}{2}$

Exercises 47
1) $-\frac{1}{4}$ 2) $-\frac{4}{15}$ 3) $(\frac{1}{6})$ 4) $-\frac{1}{8}$ 5) $(\frac{2}{5})$ 6) −2
7) 1   8) $1\frac{1}{4}$ 9) $(\frac{4}{9})$ 10) $2\frac{3}{5}$ 11) $(\frac{5}{36})$ 12) $1\frac{3}{5}$
13) $(\frac{7}{10})$ 14) 1

Exercises 48
1) $(\frac{5}{8})$ 2) $-2\frac{1}{3}$ 3) $1\frac{7}{9}$ 4) $-\frac{4}{25}$ 5) 2   6) $-\frac{2}{3}$

Exercises 49
1) −2   2) $(\frac{1}{2})$ 3) −2   4) $(\frac{1}{6})$

Exercises 50
1) $(\frac{8}{18})$ 2) $-2\frac{2}{5}$ 3) $-\frac{2}{3}$ 4) $4\frac{1}{2}$ 5) $(\frac{1}{2})$ 6) $(\frac{1}{2})$

Exercises 51
1) −1   2) $-\frac{2}{35}$ 3) −1   4) $-\frac{1}{3}$

Exercises 52
1) $10\frac{2}{3}$ 2) −3   3) $1\frac{1}{2}$ 4) 4   5) $(\frac{3}{4})$ 6) $(\frac{12}{25})$
7) $-2\frac{1}{2}$ 8) −2   9) $-1\frac{1}{2}$ 10) $-\frac{1}{2}$ 11) 3   12) $(\frac{1}{4})$
13) $(\frac{3}{4})$ 14) $-\frac{1}{7}$

Exercises 53
1) $(\frac{1}{3})$ 2) $1\frac{1}{3}$ 3) $-1\frac{1}{2}$ 4) 1   5) −1   6) $-\frac{1}{4}$

Exercises 54
1) comedies = 14, Action-adventure = 9, and drama = 5
2) 90.6   3) 5.83 inches

Self-Test, Page 71
1) B   2) D   3) B   4) C   5) D   6) A
7) A   8) B   9) A   10) D   11) D   12) D
13) B   14) A   15) B   16) B   17) A

Self-Test, Page 79
1) C   2) D   3) A   4) D   5) D   6) B
7) D   8) C   9) C   10) C

Self-Test, Page 91
1) B   2) B   3) B   4) A   5) C   6) C
7) D   8) C   9) A   10) B   11) D   12) A

Self-Test, Page 96
1) D   2) D   3) B   4) D   5) B   6) C
7) B   8) B   9) C   10) B   11) D   12) C
13) D   14) D   15) B

Self-Test, Page 108
1) C   2) C   3) D   4) D   5) C   6) A
7) B   8) D   9) A   10) A   11) D   12) A
13) C   14) A   15) B   16) D

Self-Test, Page 119
1) A   2) C   3) B   4) A   5) A   6) A
7) C   8) B   9) D   10) C   11) C   12) C
13) D   14) C   15) D   16) A   17) C   18) A
19) B

CHAPTER 4

Exercises 1
1) 3%   2) 3%   3) 36.8%   4) 71.4%   5) 18.75%   6) 89.3%
7) 84.6%   8) 47.2%   9) 26.7%   10) 112.5%

Exercises 2
1) 12.5%   2) 25%   3) 37.5%   4) 28%

Exercises 3
1) 136   2) 35.62%   3) 81.7%

Exercises 4
1) 90.5%   2) $29.40   3) $18.48   4) $13.78

Exercises 5
1) $37.80   2) $32.40   3) $54.00

Exercises 6
1) $1163.00   2) $775.33   3) $852.50   4) $664.95

Exercises 7
1) 93/250   2) 37.2%   3) 62%

Exercises 8
1) 37.5%   2) 11/40   3) 49/60

Exercises 9
1) 0.3%   2) 2348%   3) 75%   4) 102%   5) 56%   6) 7%
7) 73%   8) 201%   9) 140.1%   10) 20.51%

Exercises 10
1) 73%   2) 80%   3) 8.62%   4) 6.5148%   5) 27.5%   6) 1.6%
7) 51%   8) 6.05%

Exercises 11
1) 20.2%   2) 143%   3) 271.4%   4) 103.5%   5) 12.5%   6) 0.2%
7) 117%   8) 105%   9) 1.4%   10) 341.7%

Exercises 12
1) 13.9%   2) 86.1%   3) 18   4) 27

Exercises 13
1) 15.2%   2) $156.00   3) $188.10   4) Exceed 1.1%

Exercises 14
66.2%

Exercises 15
1) 2.31lb   2) 7.69lb   3) 42.3%   4) 83.3%

Exercises 16
1) 9/20   2) 1/4   3) 8   4) 4

Exercises 17
1) 26.4%   2) 47.1%

Exercises 18
1) $1.25   2) 45   3) 72   4) $27.50   5) $0.52   6) $30.00
7) $11.96   8) $0.125   9) $1.90   10) $13.32   11) $1.2252   12) $10.08

Exercises 19
1) $7.60   2) $75.00   3) $0.06   4) $42.00   5) $92.00   6) $0.05
7) 0.1   8) $100.00

Exercises 20
1) $0.18   2) $0.0825   3) $0.02   4) $16.74   5) $3.00   6) $1.20
7) $120.00   8) $9.00   9) 276   10) $6.00   11) $24.00   12) $15.00

Exercises 21
1) 56.25   2) 21.3   3) 45   4) 24   5) 0.32   6) 6
7) 208   8) $46.88

Exercises 22
1) 120   2) 1875   3) 375,000   4) 1388.9   5) 300   6) 825

Exercises 23
1) 270   2) $1375.00   3) $25.33   4) 102   5) 150days   6) $60.00
7) $150.00   8) 33.55   9) 10.5days   10) $33.75

Exercises 24
1) $41.39   2) $42.82

Exercises 25
1) $33.44   2) $39.60

Exercises 26
1) $2.72, $5.44   2) 12, 36

Self-Test, PAGE 128
1) D   2) C   3) C   4) B   5) C   6) A
7) D   8) C   9) A

Self-Test, page 134
1) C   2) A   3) C   4) D   5) D

Self-Test, page 141
1) D   2) A   3) D   4) C   5) B   6) C
7) C   8) B   9) D   10) D

CHAPTER 5

Exercises 1
1) Radius   2) Diameter   3) Chord   4) Radius   5) Radius   6) Center

Exercises 2
1) 106.76 ft   2) 48.356 cm   3) 18.84 in   4) 12.56 in
5) 56.52 m   6) 50.24 ft

Exercises 3
1) 5.024 ft   2) 7.8   3) 48 ft, 37.68 ft

Exercises 4
1) P = 30 cm   2) P = 12 in   3) P = 24 in   4) P = 48 ft
   A = 54 cm^2 A= 9 in^2 A = 16 in^2 A = 72 ft^2
5) P = 42 cm   6) P = 26 yd
   A = 54 cm^2 A= 44 yd^2

Exercises 5
1) 15 m   2) 5.5 ft   3) 8 cm

Exercises 6
1) Congruent angles   2) 30°   3) ∠LKM + ∠MKN   4) 122°
5) 124°

Exercises 7
a) x = 24.1 cm, P = 145 cm   b) x = 38.2 cm, P = 152.7 cm

Exercises 8
1) 12.5 cm   2) 12 ft   3) 12 ft   4) 27.50 in
5) 8 in.   6) 5 ft

Exercises 9
1) 2176 ft^2 2) 6009.96 ft^2 3) 1728 ft^2 4) 47.48 cm^2

Exercises 10
1) 4239 ft^3 2) 360 cm^3 3) 360 cm^3 4) 678.24 ft^3

Self-Test, Page 146
1) A 2) B 3) D 4) C 5) C 6) A
7) D 8) D 9) C 10) B 11) C 12) C
13) B

Self-Test, Page 150
1) C 2) A 3) B 4) C 5) A 6) B
7) B 8) C 9) C 10) B 11) B 12) B
13) D 14) C 15) B 16) C 17) C 18) C

Self-Test, Page 156
1) C 2) B 3) A 4) A 5) C 6) C
7) C 8) D 9) C 10) D 11) D 12) C
13) C 14) D 15) B 16) A 17) D

Self-Test, Page 161
1) C 2) D 3) C 4) C 5) A 6) C
7) C 8) D 9) C 10) B 11) A

Self-Test, Page 167
1) C 2) D 3) B 4) C 5) C 6) D
7) A 8) C 9) D 10) B 11) A 12) B
13) D 14) D 15) A

CHAPTER 6

Exercises 1
1) Mode = 13
   Median = 14

2) If some numbers are very higher than most often numbers in the data, then the mean will occurs higher than the median.

Exercises 2
1) Double bar graph

2) Weekly collected recyclables

Stem	Leaves
2	1, 4, 5, 5, 5, 7, 9
3	4, 9
4	4

3) Mode = 25
   Mean = 29.3
   Median = 26
   Range = 23

4) The median will also increase.

5) If the numbers are 26 to 29.3, then the mean will be decreased. If the numbers are greater than 29.3, then the mean will be increased.

Exercises 3
1) Weekly working homework

Stem	Leaves
1	75, 85, 90
2	05, 30, 30, 35, 50

2) Mode = 230
   Mean = 212.5
   Median = 217.5
   Range = 75

3) If the numbers are less than 212.5, then the mean will decrease. However, If the numbers are greater than 212.5, then the mean will increase.

Exercises 4
1) 18 2) 1/18 3) 5.6% 4) 18

Exercises 5
1) 20 2) 1/5

Exercises 6
1) 1/8 2) 6.25% 3) 3/16

Exercises 7
1) 1/18 2) 5.6% 3) 216

Exercises 8
1) 1/30 2) 3.3% 3) 30

Self-Test, Page 172
1) B 2) C 3) C 4) D 5) D 6) D
7) A

Self-Test, Page 179
1) B 2) C 3) B 4) B 5) C 6) D
7) D 8) C 9) C 10) B 11) D 12) C
13) D 14) B 15) A 16) D 17) C

Visit us at WWW.IQMATHS.com

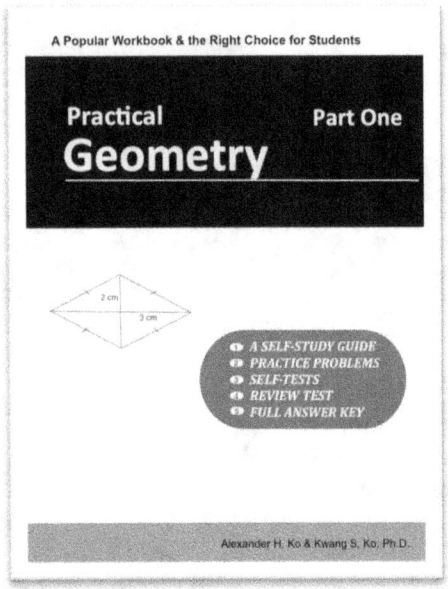

A Popular Workbook & the Right Choice for Students

Practical Geometry — Part One

- A SELF-STUDY GUIDE
- PRACTICE PROBLEMS
- SELF-TESTS
- REVIEW TEST
- FULL ANSWER KEY

Alexander H. Ko & Kwang S. Ko, Ph.D.

ISBN: 978-1-5232673-6-1

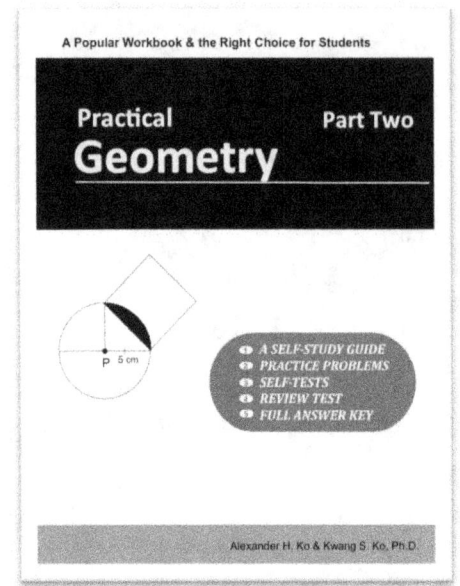

A Popular Workbook & the Right Choice for Students

Practical Geometry — Part Two

- A SELF-STUDY GUIDE
- PRACTICE PROBLEMS
- SELF-TESTS
- REVIEW TEST
- FULL ANSWER KEY

Alexander H. Ko & Kwang S. Ko, Ph.D.

ISBN: 978-1-5233620-1-1

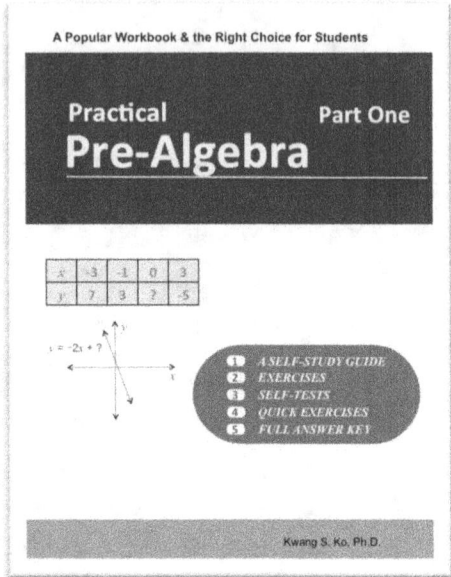

A Popular Workbook & the Right Choice for Students

Practical Pre-Algebra — Part One

- A SELF-STUDY GUIDE
- EXERCISES
- SELF-TESTS
- QUICK EXERCISES
- FULL ANSWER KEY

Kwang S. Ko, Ph.D.

ISBN: 978-1-5233628-6-8

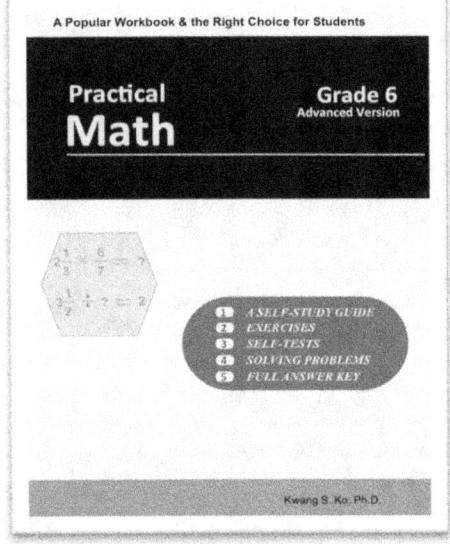

A Popular Workbook & the Right Choice for Students

Practical Math — Grade 6 Advanced Version

- A SELF-STUDY GUIDE
- EXERCISES
- SELF-TESTS
- SOLVING PROBLEMS
- FULL ANSWER KEY

Kwang S. Ko, Ph.D.

ISBN: 978-1-5233628-9-9

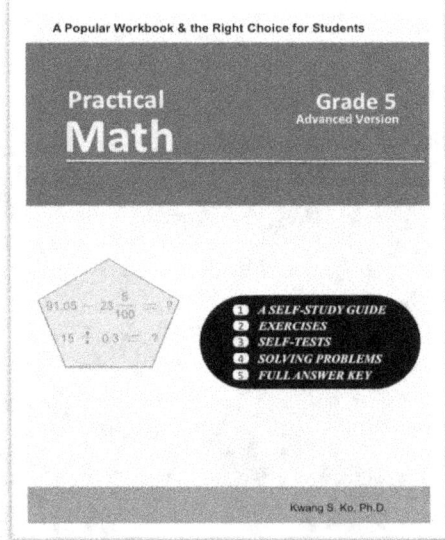

A Popular Workbook & the Right Choice for Students

Practical Math — Grade 5 Advanced Version

- A SELF-STUDY GUIDE
- EXERCISES
- SELF-TESTS
- SOLVING PROBLEMS
- FULL ANSWER KEY

Kwang S. Ko, Ph.D.

ISBN: 978-1-5233630-1-8

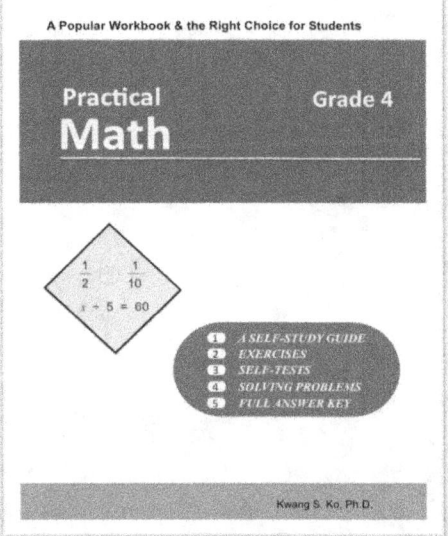

A Popular Workbook & the Right Choice for Students

Practical Math — Grade 4

- A SELF-STUDY GUIDE
- EXERCISES
- SELF-TESTS
- SOLVING PROBLEMS
- FULL ANSWER KEY

Kwang S. Ko, Ph.D.

ISBN: 978-1-5233630-2-5

Other books are sold at WWW.IQMATHS.com.